Decolonising Schools in South Africa

This book explores the challenge of dismantling colonial schooling and how entangled power relations of the past have lingered in post-apartheid South Africa.

It examines the 'on the ground' history of colonialism from the vantage point of a small town in the Karoo region, showing how patterns of possession and dispossession have played out in the municipality and schools. Using the strong political and ontological critique of decoloniality theories, the book demonstrates the ways in which government interventions over many years have allowed colonial relations and the construction of racialised differences to linger in new forms, including unequal access to schooling. Written in an accessible style, the book considers how the dream of decolonial schooling might be realised, from the vantage point of research on the margins. This Karoo region also offers an interesting case study as the site where the world's largest radio telescope was recently located and highlights the contrasting logics of international 'big science' and local development needs.

This book will be of interest to academics and scholars in the education field as well as to social geographers, sociologists, human geographers, historians, and policy makers.

Pam Christie is Professor Emeritus in Education, University of Cape Town, and Honorary Professor, University of Queensland. As well as university research and teaching, she has worked with schools, government departments, and NGOs on educational change and social justice.

Routledge Research on Decoloniality and New Postcolonialisms

Series Editor: Mark Jackson, Senior Lecturer in Postcolonial Geographies, School of Geographical Sciences, University of Bristol, UK.

Routledge Research on Decoloniality and New Postcolonialisms is a forum for original, critical research into the histories, legacies, and life-worlds of modern colonialism, postcolonialism, and contemporary coloniality. It analyses efforts to decolonise dominant and damaging forms of thinking and practice, and identifies, from around the world, diverse perspectives that encourage living and flourishing differently. Once the purview of a postcolonial studies informed by the cultural turn's important focus on identity, language, text and representation, today's resurgent critiques of coloniality are also increasingly informed, across the humanities and social sciences, by a host of new influences and continuing insights for different futures: indigeneity, critical race theory, relational ecologies, critical semiotics, posthumanisms, ontology, affect, feminist standpoints, creative methodologies, post-development, critical pedagogies, intercultural activisms, place-based knowledges, and much else. The series welcomes a range of contributions from socially engaged intellectuals, theoretical scholars, empirical analysts, and critical practitioners whose work attends, and commits, to newly rigorous analyses of alternative proposals for understanding life and living well on our increasingly damaged earth.

This series is aimed at upper-level undergraduates, research students and academics, appealing to scholars from a range of academic fields including human geography, sociology, politics and broader interdisciplinary fields of social sciences, arts and humanities.

Making Urban Theory
Learning and Unlearning through Southern Cities
Mary Lawhon

Decolonising Schools in South Africa
The Impossible Dream?
Pam Christie

For more information about this series, please visit: https://www.routledge.com/Routledge-Research-in-New-Postcolonialisms/book-series/RRNP

Decolonising Schools in South Africa

The Impossible Dream?

Pam Christie

Routledge
Taylor & Francis Group
LONDON AND NEW YORK

First published 2020
by Routledge
2 Park Square, Milton Park, Abingdon, Oxon OX14 4RN

and by Routledge
52 Vanderbilt Avenue, New York, NY 10017

Routledge is an imprint of the Taylor & Francis Group, an informa business

British Library Cataloguing-in-Publication Data
A catalogue record for this book is available from the British Library

Library of Congress Cataloging-in-Publication Data
A catalog record has been requested for this book

ISBN: 978-0-367-42575-3 (hbk)
ISBN: 978-0-367-85362-4 (ebk)

Typeset in Times New Roman
by codeMantra

For Dawn

Contents

Figures

Maps

Tables

Acknowledgements

Thank you, first and foremost, to my research companions on this project, Dawn Butler and Sanet Lombard, and to Heather Jacklin, Carolyn McKinney, and Mignonne Breier for their support and encouragement. Sincere thanks to Barry Firth for opening doors for me; to Sakkie Potgieter for sharing his vision; to Louise Boezak, Bé Potgieter, and Heinrich Jansen for their many insights on life in Carnarvon; and to Pieter and Rebecca at the Lord Carnarvon Hotel for their hospitality. Thanks also to Kai Barron, Sara Black, and Christopher Hedemann for assistance with research technicalities, and to Brenda Cooper for the schtick. For intellectual companionship, my gratitude to the 'Shut up and Write' group in the School of Education at University of Cape Town and my co-teachers Kate Angier and Rochelle Kapp, who read parts of the draft text. For another view on the world, thanks to Neil Rusch, Jose Manuel de Prada-Samper, and John Parkington, and for a different view again, to Affrica Taylor. For administrative support, thanks to Thembakazi Rala; for library assistance to Loreen Rushby in Government Publications at the University of Cape Town; and for library access to the University of Queensland School of Education. The research for this book could not have been completed without the generous financial support of South Africa's National Research Foundation, which I gratefully acknowledge. My thanks, finally but crucially, to the people in the schools in Carnarvon who tolerated an inquisitive outsider and whose experiences are included in this book.

1 The long reach of coloniality

Setting the scene from a marginal place

The place of Carnarvon

In South Africa's semi-arid and remote Karoo region – an interior plain of vast skies and stretched out land – lies the small town of Carnarvon. The Karoo is one of the world's richest palaeontological sites containing fossils of earliest life forms, and it also holds a multitude of archaeological scatterings showing hominid presence from more than a million years ago. Archaeological and historical research attests to indigenous hunter-gatherer presence on the land, and movements of nomadic pastoralists across it, hundreds of years before European occupation. Colonisation brought fierce competition over sparse resources of grazing land and water, violent confrontations, decimation of indigenous people, and distinctive patterns of settler colonialism in waves of possession and dispossession. Carnarvon is one of several small towns established in the Karoo in the 1800s, and its present-day conditions provide a particular vantage point for tracing the development of racial capitalism in South Africa over time. Embedded inequalities echo from the past to the present in this marginal place and are reflected also in schooling arrangements and outcomes. The current circumstances of the town highlight the difficulties of achieving social and economic change in spite of South Africa's political transformation of 1994.

Because of its pristine skies, stable geological formation and remote location, this place has been selected for the construction of South Africa's share of the Square Kilometre Array (SKA), an enormously costly international project in 'big science' (on the scale of the European Organisation for Nuclear Research, CERN). Farms near Carnarvon have been turned into the site of the world's largest radio telescope array probing the origins of the universe. This place, where there are traces of the oldest forms of life on earth, is thus a focal point for a cutting-edge future-focused endeavour in astronomy. Remote and semi-arid as it is, however, this is a peopled place. The occupation by the SKA invites the interpretation of yet another wave of displacement, this time of farmers and others living in a rural service centre being displaced by the infrastructure of big science. Justifying this very costly endeavour, the post-apartheid government has asserted the importance of expanding scientific knowledge alongside the provision of basic needs in what is a deeply unequal society.

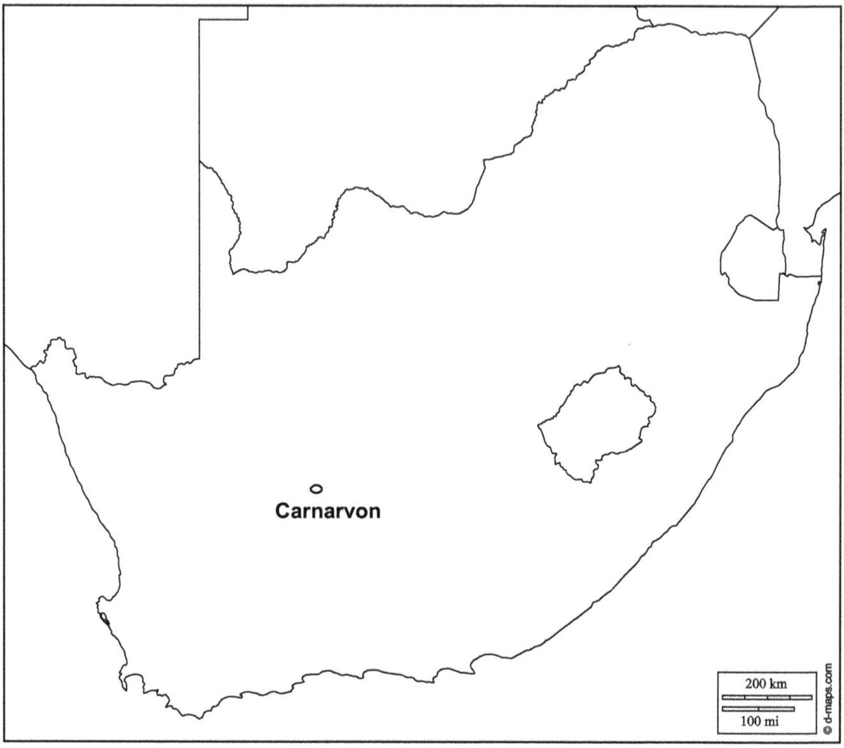

Figure 1.1 Approximate location of Carnarvon, South Africa.

Place and schooling

Schooling in the town of Carnarvon in the Karoo provides the spine of this book, running from colonial times to the period after the end of apartheid which offered a 'decolonial' moment for South Africa.[1] The study of schooling, place, and time reflects my long-standing commitment to working for social justice in education. Schools are emblems of modernity, and though they are flawed institutions, their social mandate is important: to provide systematic teaching and learning for young people in ways that prepare them to contribute to a shared world and, hopefully, change it for the better. Schools do not always fulfil this mandate, and I believe it is important to grapple with why this is so and how to improve the experiences and learning outcomes of students to achieve greater equity.

Having visited hundreds of schools in South Africa and Australia over many years as a researcher and teacher educator, I have been fascinated by the differences as well as similarities between schools. There is a 'sameness' to schools, such that it would be hard for me to come upon a school and not recognise it for what it is – even if there are no buildings, or if classes are conducted under trees. At the same time, despite their sameness, schools are very different places, and each has its own distinctive vitality or tonality.

In general, though, it is important not to romanticise differences between schools, since these differences also express the profound inequalities of their broader societies. Indeed, there is general agreement among sociologists of education that schools and classrooms enact inequalities in multiple ways – though there is fierce debate about how to explain and remedy these practices. By now, there is a powerful accumulation of research showing that inequalities in schooling follow predictable patterns of social class, race, gender, region, and language. Research also shows the stubborn endurance of inequalities in the face of multiple attempts to shift them.

Where once inequalities in schooling in South Africa could be straightforwardly related to apartheid (although not reduced only to this), the persistent inequalities in post-apartheid schooling have perplexed policy makers whose intention has been to redesign the system towards greater equity and quality. In addition to the imprint of apartheid, remnants of interventions that have failed to achieve what they intended to are also visible. Understanding the persistence of deep problems in the face of policies designed to achieve equity and excellence is an important task to grapple with.

The schools in the town of Carnarvon are the centre point of the empirical work of this study, stretching from colonial times to present-day times of 'big science'. From the vantage point of this marginal town, the book throws light on larger patterns of schooling, place, and the social formations they fold into, over a span of time. In magnifying one place on a very large canvas of schools, the intention is to provide insights into the multiple ways in which inequalities in schooling take shape, endure, and shift – insights that are not easily identified by the broad histories and policy analyses of education alongside which this book is written. From the vantage point of this marginal town in the middle of South Africa, the book begins by laying out the palimpsest of colonial schooling and examines its endurance in post-apartheid times, with the aim of stretching beyond current ways of understanding continuity and change in schooling. Drawing on the analytical terrain opened up by theories of decoloniality, the book explores how historical inequalities linger in South Africa after the seismic political changes that heralded a more equal social order. It explores what might be required to shift these inequalities in current global contexts, towards an ethical consideration of how we might live together with all others in the world we share.

Colonialism and apartheid

During the period that I was visiting Carnarvon for fieldwork from my base at the University of Cape Town (UCT), student protests erupted at this university and others. Initially a stand against the prominently positioned statue of Cecil John Rhodes, a master-capitalist of colonialism, the UCT protests soon broadened from #RhodesMustFall to the nationwide call for #FeesMustFall. In the course of protests, students highlighted issues of 'coloniality', including racism, the Western-oriented curriculum, and cultural exclusion. Before long, school student protests hit the media, with black girls in particular complaining that hair policies were discriminatory in former white, now integrated, schools, and that they were penalised for speaking their home languages at school. In short, black students challenged the ethos of schools that had changed very little despite the raft of new education policies which enabled racial desegregation. And this is to say nothing of conditions in the majority of schools in South Africa, located in former townships and rural areas, which have remained predominantly black and comparatively under-resourced and are generally the worst performing in the system – surely a remnant of apartheid and colonialism.

The challenges raised by students led colleagues and me to explore the literature on decoloniality, in part at least to understand why it captured their imagination so powerfully (Christie & McKinney, 2017). As one who had felt uncertain about using theories of postcolonialism to analyse South African conditions, I found that de/coloniality theories opened useful theoretical spaces for exploring the particular conundrum facing South Africa and its education system: continuing economic, social, cultural, and spatial inequalities remaining so starkly evident long after the massive political changes of 1994.[2] To give specific examples: the burdens of poverty and unemployment continue to be skewed towards black people; 'race' has certainly not disappeared as the major predictor of social and schooling outcomes; African languages are given almost no space in the school curriculum; the schooling system systematically fails the majority of students (who are black); and the racial geography of apartheid remains all too evident in practices of daily life including schooling. (I substantiate these points in later chapters of this book.)

In accounts of colonialism and decolonisation, the end of apartheid is rightly taken to be a watershed moment. Yet South Africa's association with colonialism (and postcolonial theory) is not a straightforward narrative and warrants some elaboration at this point. South Africa has a complex and violent history of colonialism long predating apartheid and its history is not amenable to easy labelling. Briefly, white settlement at the Cape began with a Dutch trading station in 1652, taken over by the British in 1806, and including waves of other European settlers (French, German, Scottish). European pastoralist expansion into the region over the next century was achieved through violent contestation over land and resources, resulting finally in the subjugation of existing inhabitants. In the early phase of encounter, local indigenous KhoeKhoe (herders) and San (hunter gatherers) people were largely decimated, with remnant groups displaced into desert areas in the interior. Bitter battles were waged to defeat the significant African polities living across the region. Africans were successively dispossessed of their land and proletarianised, becoming labourers on farms, and later on in mines and in manufacturing as the economy developed. At various times, labour was imported – slaves from several sources including elsewhere in Africa as well as India and South-East Asia to work in the Cape settlement, labourers from India for sugar cane plantations in Natal, and indentured Chinese labourers for mining on the Witwatersrand – and this added to the diversity of the population. European penetration into the interior resulted in four different settler polities by the end of the 1800s: two differently governed British colonies and two fragile Boer Republics. In these settler polities, schooling was often meagre, and schooling established by British and Dutch settlers expressed their different aspirations and worldviews. Schooling for black people was left entirely to missionary endeavours. The discovery of minerals – diamonds and then gold – was decisive in the economic development of the country, catalysing British interest in the Boer-held goldfields and leading to the Anglo Boer War (alternatively named the South African War) of 1899–1902.

In the initial colonial period, racial and gender hierarchies and different forms of labour control were put in place in coercive ways, well before apartheid institutionalised these.[3] It is important to recognise, though, that colonial rule – whatever its forms – was always contested, with original occupants far from passive in their responses, and hegemony never stably attained by successive governments.

In 1910, the Union of South Africa was formed as a British dominion and much later, in 1960, South Africa declared itself a republic and left the British Commonwealth. Both of these are moments relating to colonialism, rendering complexity to notions of 'postcolonialism' in temporal terms. The establishment of the Union formalised racial and spatial inequalities. The limited land and franchise rights of black people were reduced. The 1913 Land Act (adjusted in 1936) confirmed the division of land achieved by settler conquest, enshrining white ownership of 80% of the land and confining black land ownership to reserve areas under government-endorsed traditional leaders. In response to these measures, the African National Congress was formed in 1912, sending a delegation to negotiate – unsuccessfully – with the King of England. In the ensuing period, segregation was entrenched along with the development of capitalism. In the 1920s, 'job reservation' legislation secured employment for whites at the expense of people of other races, and the ownership of wealth and productive forces was held exclusively in the hands of whites. Black access to urban areas and possibilities for land ownership were curtailed. Public schooling for whites became progressively free and compulsory, but racial segregation in schooling remained, with the state making no provision for black people beyond subsidies for mission schools. Thus, by the time the Afrikaner nationalist party came to power in 1948 with its ideology of apartheid, the groundwork for spatialised relations of power linked to labour control and capitalist development were in place, with space already linked to race, and access to opportunities already linked to race and place (Christie, 2013).

While there is no doubt that apartheid was the apogee of centuries of violent conquest, there has nonetheless been considerable scholarly debate about the form and nature of the apartheid state. Without 'reinventing the wheel', three examples will suffice for an audience unfamiliar with Soth African studies. First, the particular configuration of race and class in capitalist development in South Africa has been much debated (see e.g. Posel, 1997; Dubow, 1989; Wolpe, 1972). A central issue has been to understand the nature of the relationship (e.g., correlative, causal, or contradictory) between racial segregation and capitalist development, particularly as this has played out in varied forms of labour control, spatial divisions, and measures to control the movements of black people.

Second, in relation to education, issues of race, class, and the apartheid state have also been much debated. A particular focus has been on the early move of the apartheid state to close missionary schools and bring education for black people under tight state control – the infamous 'Bantu Education' system of segregated and unequal schooling. Understanding the interplay of the different logics of racial segregation and labour reproduction in the establishment of apartheid education has been vigorously debated (see Kallaway, 1984; Christie & Collins, 1982).

A third example surrounds the particular forms of spatial division of land. In the 1960s and 1970s, the state established ethnically based 'bantustans' in former reserve areas, and afforded limited rights to African people, classified according to ethic groups, in their designated bantustans – while denying them South African citizenship. Land allocation policies meant that the bulk of the country's productive resources (minerals and most of the farming land) were secured in white hands. Bantustans were spatially splintered, too over-populated to sustain productive agriculture, and bypassed by networks of transport and communication. During the 1960s and 1970s, bantustans, administered by government-paid chiefs and officials, were steered towards forms of self-government. Between 1976 and 1981, four of the ten bantustans were granted 'independent' status by the South African government (though they were not internationally recognised) – a complex variant of 'decolonisation' at the time when much of Africa was being decolonised.

In terms of colonialism, the apartheid state has been variously theorised, for example, as 'internal colonialism', 'colonialism of a special type', 'settler colonialism', and so on (see Evans, 2012; Everatt, 1992; Hopkins, 2008; Mamdani, 1996; Saunders, 2000). These are not simply semantic or even analytical differences in a complex theoretical terrain; they also point to differences in political struggle at particular moments.[4] This means that the terminology associated with colonialism needs to be used with care in South Africa's case. And while the end of apartheid may indeed be viewed as an end to the colonial era, simplifying this into a semantic designation may gloss over a deeper understanding of complex historical processes explored in specialist scholarship. Similar problems apply when postcolonial or de/coloniality theorisations are applied too literally to South African conditions, without taking historical specificity into account – to say nothing of local specificity.

Without further discussion of South Africa and colonisation, the approach taken here is that theories of postcolonialism, de/coloniality, and 'southern theory' may be understood as different entry points in mapping a shifting analytical terrain that is simultaneously marked *both* by histories of colonialism *and* by changing contextual issues at a global level. The current context has generated a particular set of analytical challenges, including global neoliberalism, cybertechnologies, shifting identities and inequalities, climate change, ecological damage, and global health pandemics. Spanning these different strands of debate, I suggest, is a concern for developing new ways of living in the world beyond existing ontologies of difference and including the more-than-human world.

Taking the vantage point of a marginal town like Carnarvon provides a context for exploring these intertwined issues and the particular challenges facing South Africa. The enormously costly Square Kilometre Array telescopes in the locality of a poor and marginal town stand as signifier of the unresolved relationship between 'big science' and human development needs. And in the penumbra of this venture in international astronomy lie schools that offer limited opportunities to students in the poor and divided communities they serve.

Decolonising schooling?

On the shifting analytical terrain that I have mentioned, educationists face particular challenges given that our work lies both in scholarly understanding and professional practice. On this terrain, the larger project of this study grapples with several key issues that are highlighted with particular clarity by theories of de/coloniality and have direct application to education. These raise questions such as: how should we understand the enduring, intersectional inequalities that remain after the formal demise of colonialism, and how do we work against these? What is required epistemologically for working *within* the 'border' contexts we find ourselves in? How do we shift the form of current conversations towards a different ethics of engagement with all others in the world we share? I look briefly at each of these three questions in turn, using de/coloniality as a springboard.

First, in grappling with the inequalities that continue after colonial administrations have formally ended, it is useful to name these inequalities in a way that differentiates them from colonialism itself. Demarcating the difference between colonialism and coloniality accords a particular space to explore the nature of the latter, explained as follows by Nelson Maldonado Torres (2007):

> Coloniality is different from colonialism. Colonialism denotes a political and economic relation in which the sovereignty of a nation or a people rests on the power of another nation, which makes such nation an empire. Coloniality, instead, refers to long-standing patterns of

> power that emerged as a result of colonialism, but that define culture, labor, intersubjective relations, and knowledge production well beyond the strict limits of colonial administrations. Thus, coloniality survives colonialism. It is maintained alive in books, in the criteria for academic performance, in cultural patterns, in common sense, in the self-image of peoples, in aspirations of self, and so many other aspects of our modern experience. In a way, as modern subjects we breathe coloniality all the time and everyday.
>
> (p. 243)

In relation to how lingering inequalities may be remedied, theorists of de/coloniality firmly close the door on easy assumptions that all previously colonised countries could proceed along the path of Western modernity. Linking the enlightenment and the emergence of Western modernity to the fifteenth century Atlantic trade circuits (including slavery) and the colonisation of the Americas by Europeans, de/colonial theorists argue that Western modernity is inextricably linked to colonialism in a co-constitutive way. Western modernity took shape in the economic, social, and epistemological relationships of colonialism – conditions which are not straightforwardly available for former colonies in the form of 'development'. In other words, Western modernity, imbricated as it is in the economic, social, and epistemological relationships of colonialism, is not an option that is straightforwardly available for former colonies in the form of 'development'. 'Progress' in the name of modernity is an illusion.

If this is so, the implications for schooling in South Africa are immense. It questions the common assumption that historically disadvantaged schools are on the developmental path to becoming the same as their 'modern', historically advantaged counterparts. If this cannot be assumed, what options might there be for reducing inequalities in post-apartheid schooling? This is a troubling question that I return to throughout the chapters of this book, taking the vantage point of a marginal place to highlight what this might entail.

The second key issue concerns the adequacy of describing and conceiving of the world from universalising Western perspectives alone. Recognising the power relations of knowledge, de/coloniality theorists argue that these particular universalist perspectives have been historically imposed on other parts of the world that are then ranked as inferior. Recognising the situated nature of knowledge, these theorists argue instead for an epistemic approach of pluriversalism. In saying this, a crucial point is that this does not mean that the Western episteme should be discarded or be supplanted by indigenous knowledge or precolonial knowledge systems. Nor does it mean cultural relativism. What is required, rather, is to acknowledge and work within the entanglement of knowledge systems, and to strive for different ways of thinking and understanding from within the situated complexity of border positions. In other words, what is required is to think from *within* the border, rather than *about* the border, in working against inequalities to find sustainable alternatives. This requires an openness to exploring different forms of ethical relationship in the world we share with others, human and other-than-human.

In South Africa, protests by university and school students highlight the complexities of working within border conditions of knowledge. Given the association of 'excellence' and 'powerful knowledge' with the canonical and disciplinary knowledge of the Western episteme, it is important that this not be simply set aside in a gesture of protest. At the same time, the historical and situated nature of this canon needs to be recognised, and questions raised about its sufficiency and adequacy in understanding the situated complexities of a country like South Africa. Also to be recognised is the violence that is implicit when languages other than English are devalued as lesser or regarded as irrelevant. These issues have no simple solutions, requiring instead a willingness to work with them as tensions inherent in border contexts.

The third key issue is how to move towards a different conversation, beyond the inevitabilities of Western modernity and working towards a different, pluriversal imaginary. As Walter Mignolo (2007) states:

> Decoloniality, then, means working toward a vision of human life that is not dependent upon or structured by the forced imposition of one ideal of society over those that differ, which is what modernity/coloniality does and, hence, where decolonization of the mind should begin. The struggle is for changing the terms in addition to the content of the conversation.
>
> (p. 459)

On the analytical terrain I outlined earlier, there is a global need for ethically reframing how we might best live together in a common world we share with all, human and other-than-human. This is an enormous challenge for those working in schools, which are unbending as institutions of modernity, yet it is one that must be grappled with. In the conclusion to this book, I address the issue more fully, in discussion with the 'one world' position advocated by Achille Mbembe and others and described by Mbembe as the need for 'repair'.

Research from the vantage point of Carnarvon

Using the broad framework of de/coloniality, this study takes the provision of schooling as its centre point, with the aim of investigating how inequalities take shape, endure, and change over time. It presents the findings of research on a single place and its schools, using a place-based study as a vantage point for insights to accompany broader historical studies and policy analyses of schooling. The study focuses on two time periods: the establishment of the town of Carnarvon and its schools as part of processes of colonisation, and the contestation over the schools in the post-apartheid period as an indicator of the complexity of changes away from the power relations of colonialism. This complexity illustrates the inequalities of colonialism that linger on after changes in government – conditions of coloniality that require further work to achieve more fundamental change. The study draws on the rich historiography of the Karoo and South Africa more generally, on extensive literature on education in South Africa and more broadly, on literature on the political transition of 1994, and on my own mixed method research findings.

The value of taking schooling as the focal point is two-fold: schooling is an important institution worthy of study in its own right; and it is closely related to broader social practices and imaginaries as they emerge and change over time. Schooling as a social institution has accompanied industrialisation in the West and colonisation in Africa and elsewhere. Schooling systems were instituted in virtually all European states in the nineteenth and early twentieth century, were prominent institutions of colonisation, and were signals of modernity in newly independent African countries (see Fuller, 1991). The value of taking a single place as the focal point in a broader narrative is that it is possible to explore the textures and complexities that are necessarily glossed over in general policy studies and histories. Including the dimension of history enables closer consideration of continuities and changes over time. Building on the large body of national and international research showing well-established patterns of inequality in schooling, the intention of this study is to provide a more fine-grained understanding than is possible through statistical analyses or policy studies alone.

The premise of this book is that the study of schooling in place is able to show in nuanced ways how the power relations of coloniality take enduring forms. Particular patterns of possession and dispossession become evident upon closer study, and it is these very patterns – often in their daily and micro enactments – that must be understood and engaged with if policies and practices are to shift. Locating an in-depth study within a larger analysis of policy and history shows the fissures and gaps in governmentality, and also the stark ways in which policies fall short in their ambit claims of change. Without detailed engagement at this level, theories of change – including postcolonialism and decoloniality – cannot come to grips sufficiently with what may be entailed in shifting beyond current inequalities and exclusions, how they are lived and resisted, and how change may be envisaged and enacted.

The fieldwork research of this study combines document study, systematic observations, and formal interviews, with views and information gathered more informally from people and place. In drawing together an account of the schools in the town based on this range of resources, I have taken care to leave an 'audit trail' as a form of research accountability. My intention is to provide a trustworthy account that contributes to scholarship, but not to 'neaten' the untidy mix too much.[5] Too much distillation in the quest for clarity may smooth over the complexities and contradictions in what Henri Lefebvre (1991) terms 'the encounter of everything' (p. 101) – although too little structure may fail to show the patterns and regularities in daily life.

I selected Carnarvon for my research as a distinctive yet marginal place, anticipating that a study of schools from the vantage point of this place would give a specific face to well-established social patterns of schooling and how its inequalities might be shifted – it would give a vantage point to magnify 'difference' in the generalisable 'sameness' of these patterns. I had anticipated that, armed with my research design and university ethical clearance, I would be able to investigate schooling arrangements, how these related to the comings and goings of daily life of the town, and what schooling offered for the future of young people living in this place. I had also hoped that it would be possible to throw some light on the links between the 'big science' of the SKA and improvements for local people, particularly in schooling. However, I had misread the complexity of what I was to find.

On my first visit to Carnarvon, one of the most striking features of the two schools in the town was the total absence of white students, though the town was clearly dominated by the resources of white people living in the town itself and on surrounding farms. How could this be, so long after the official end of apartheid? Although well aware that research on schooling in South Africa cannot avoid dealing with matters of 'race', I was nonetheless surprised by what seemed to be an extreme divide in a small place serviced by few schools. I had anticipated that the schools' results would be influenced by language differences among students (as is the case in many South African schools), only to find that all of the students and teachers were Afrikaans-speaking and that language of instruction was not a barrier to learning. My prior scanning of the Senior Certificate and Annual National Assessment results for the two schools (results which are generally mediocre to poor) did not prepare me for the classrooms I saw – which were well-ordered and mostly staffed by an appropriately qualified mix of Afrikaans-speaking teachers.

Given that very little has been written about this remote and marginal place, I had come to the town with no knowledge of the history of contestation over the control of its schools in the post-apartheid period, since this is not recorded in available secondary sources. I had no idea that the former white school had fought bitterly against attempts by the new Northern Cape Provincial Government to restructure the schools in the town in 1997, and I certainly did not anticipate the visceral anger expressed by many white interviewees about the amalgamation of schools ten years later. I also had no idea of the significant role played by the church in the history of the town, or that the town's Uniting Reformed Church was involved in the international reformed church movement that declared apartheid to be a heresy in 1982. I did not know that the town had been active as a rural node in the political struggle against apartheid in the 1980s. I could not have anticipated that the pastor, who had faithfully served the local Coloured church in this remote town for more than twenty years, would be a white, middle-aged Afrikaans-speaking missiologist who had studied Marxism and theology in Germany for his doctorate and was himself politically active in giving voice to the poor. (I observed his wife, also a Stellenbosch graduate, teaching with great care and commitment in her visually colourful classrooms in the Carnarvon primary school on my research visits.) In short, in selecting this small, remote Karoo town as a place to study, I was unaware that it had a very particular history, whose reverberations would continue to be felt in the town to this day.

Needless to say, my research endeavours became far more complex than I planned, stretching me far beyond my comfort zone. As an English-speaking visitor, I found myself an immediate outsider in this Afrikaans-speaking town, regarded with suspicion as I asked questions and probed issues. It took many more visits than I had planned for people to talk to me about the schools, which they did with enormous reluctance. Without the assistance of a colleague who had taught in the town in the 1990s, and to whom I owe an enduring debt of gratitude, it is unlikely that doors would ever have opened for me as they did. I found myself struggling on many an ethical knife-edge, not least in relation to university ethics requirements for 'informed consent' for structured interviews and observations, but also because of the detail of some of what I saw and heard. In my research process, I conducted formal interviews and I also spoke informally to people in the town and the schools. I had permission to observe classes in the schools, and I spent many hours on informal observations around the town. I walked and drove the streets of the town and its surroundings over the course of five years, making short visits during different seasons and school terms. I visited the Northern Cape Education Department in Kimberley, and the District Offices in De Aar, as well as the SKA offices in Cape Town and Johannesburg to gain information about the broader context of the schools in Carnarvon and the programmes and people that supported their endeavours. And in between my visits I scoured archives and libraries to find out what I could about the history of the town.

Finding documentary sources proved unexpectedly elusive. In the Cape Archives, I found that records of certain times and events were simply missing, with the archived records for Carnarvon's Coloured schools signing off abruptly in November 1989 with the note 'File Closed'. Records of the Rhenish Missionaries for this period are housed in Wuppertal in Germany, and the sample available to me in Cape Town are written in a Gothic German which I could not read. From what I could gather, these records also provided very little on the Carnarvon mission station and its school. The original records held by the Carnarvon mission church were taken away by an earlier researcher and are hopefully still in existence somewhere that I have not been able to locate. No official investigation was conducted when the primary school burnt down in 1989, and no one I interviewed could say conclusively what had happened. Nor could I piece together a satisfactory picture of the intense internal dynamics of the two Coloured schools that operated in completely unsuitable, overcrowded conditions on a single set of premises for nearly twenty years after the primary school was relocated there without being rebuilt after the 1989 fire. With regard to the former white school, *Hoërskool* Carnarvon, litigation between its governing body and the Department of Education was not always straightforward to locate, with some cases being joined to other schools' litigation. Carnarvon is too small to have a newspaper of its own, and the regional newspaper, Calvinia's *Noordwester*, provides sporadic coverage of events in the town.

The result, for me, was an abiding sense that just beyond my grasp lay the evidence I earnestly sought to build a more precise account of schooling in this place. I felt a lingering concern that I was observing a picture through reflections in the glass, not as clear or sharp as I would like.

As I wrestled with the discomfort I experienced as an outsider in a place of evident inequality and social complexity, I found it necessary to reflect more carefully on the social relations of my research: the power relations of 'researcher' and 'researched'; the objectification of others in the research process and its 'outputs'; the resistances I encountered from people who were unwilling to participate in research that they did not support on a contested political issue; the danger that, in a context of asymmetrical power relations and value-differences, my 'interviews' could be perceived as 'interrogations', and 'home visits' as 'invasions of privacy', thereby evoking unhelpful resistance and evasions.

Consequently, while reaffirming my research aims, I approached my research process with greater attention to its power relations and sensitivities. Since I regard research itself as an artefact, albeit following accountable procedures and conventions, my account would need to assemble what was available to me, acknowledging the silences, absences, and blocks as part of the yield. My task would be to work respectfully with what was available, without attempting to hide the gaps or shape the fragments into a neat and complete narrative that did not fit what I had experienced. Through discussions with colleagues about the ethical and methodological dilemmas of this task, I found the work of Linda Tuhiwai Smith (2012) on decolonising methodologies and Mary Louise Pratt's (2007) notion of the 'contact zone' to be particularly helpful, since both give particular consideration to power relations of research in colonised contexts. As explained by Pratt, the 'contact zone' refers to a space of encounter where the 'centre of gravity and the point of view' are shifted, particularly when power relations are unequal as in colonial encounters.[6] Pratt writes of the contact zone as follows:

> It invokes the space and time where subjects previously separated by geography and history are co-present, the point at which their trajectories now intersect. The term "contact" foregrounds the interactive, improvisational dimensions of imperial encounters so easily ignored or

> suppressed by accounts of conquest and domination told from the invader's perspective. A "contact" perspective emphasizes how subjects get constituted in and by their relations to each other. It treats the relations among colonizers and colonized, or travelers and "travelees," not in terms of separateness, but in terms of co-presence, interaction, interlocking understandings and practices, and often within radically asymmetrical relations of power.
>
> (p. 9)

Viewing my fieldwork research in Carnarvon as a 'contact zone' activity enabled me to foreground the co-presence of myself as a researcher with others in the town, interacting in the materiality of place, with representations of it in documents and photographs (my own and those of others), and acknowledging the lived interactions of meaning-making in the research activities. Approaching my research process as working with co-presence in complex interaction, I crafted a hybrid account of everything I could gather together, including what I had experienced as resistance or partial truths, and as elusive, blurred, or missing information, to understand how particular patterns of inequality in schooling were formed in this place, and how they have endured while also shifting over time.

The notion of the contact zone of research also meshed with my intention to think with Lefebvre's notion of social space as an 'encounter of everything' – a point I explain more fully in the following chapter, where I outline the various theories I draw on as a framework for my analysis of schools, opportunity, space, and place.

In this spirit, through the chapters on the first time period addressed in the book, I have used mostly historical sources (largely secondary but also primary) to show the impact of colonial expansion into the semi-arid, sparsely peopled land in the 1800s, and to show how patterns of possession and dispossession took racialised forms that hardened over time. I show that this interpenetration of race and class manifested not only in the ownership of land and economic resources, but also in the town's institutions, in particular the municipality, church, and school. I show that in this specific colonial context, stretching far away from the government in Cape Town, schools were established through the efforts of different social actors with their own intentions and aspirations, and during this period the foundations were laid for segregated and unequal schooling. I argue that the social relations established in these colonial times have continued to reverberate through the period of apartheid and into the period beyond its formal demise, and that they have salience in the current ownership of land and economic resources, as well as in the town's institutions. The struggle over the control of these institutions – in particular the schools – has been a central dynamic in the social relations of the town, exemplifying the intersectional relationships of lingering inequalities characteristic of coloniality.

For the second time period of the book focusing on the post-apartheid period, I have placed the struggle over the schools in Carnarvon in the context of post-apartheid education restructuring. The place-specific picture I present in these chapters draws on my approach to 'research-as-contact zone'. I use the vantage point of this specific place to show the compromise nature of the 1994 political settlement and the limited extent of redistribution in schooling and more broadly – insights into conditions of coloniality. These chapters show how the ambitious and ambiguous system of school governance enabled individual schools to resist the equalising intentions of state policies. They also show the cross-cutting effects of policies on curriculum, assessment, and inclusion as they are enacted in schools in conditions of structural poverty. They point to the bare provisioning of schools that rely on state funding alone, and the limited opportunities that these schools are able to afford to the young people they serve. And they illustrate the social imaginary of difference that needs to be challenged if the decolonisation of schooling in South Africa is to be more than an impossible dream.

The penultimate chapter of the book looks at the arrival of the SKA into the vicinity of Carnarvon, exploring its potential to shake the sedimented social and economic relationships in ways that the political changes of 1994 could not. This is a possibility that is perhaps visible – if only just – on the very horizons of this study. The chapter considers the interaction between 'big science' and 'development', both from the perspective of this particular place and in consideration of broader issues of skills and capacity-building in science in South Africa.

In the concluding chapter, I explore the possibilities of shifting the inequalities in schooling, picking up on the challenge of decoloniality theorists and others that the form of the conversation needs to be changed as well as its contents. In doing so, I propose the need for alternative ethical considerations about how we might live together in the world we share with all others.

Notes

1 This book is based on a research project entitled Space, Place and the Affordances of Schooling in South Africa, funded by South Africa's National Research Foundation from 2015 to 2019.

2 For a thorough problematization of the term 'postcolonial' in the context of colonization in Africa, see Ngũgĩ wa Thiong'o (2012). 'Decoloniality' refers to the work of a predominantly Latin American group including Mignolo (2007) and Maldonado Torres (2007).

3 Much is written on this. For a general history, see Hamilton, Mbenga and Ross (2012) and Ross, Mager, and Nasson (2012).

4 This difference relates in particular to the political positionings of the ANC and South African Communist Party, as set out in Everatt (1992).

5 I undertook six short research visits to the town and its schools, conducted more than 30 formal interviews, and had many dozens of informal conversations and periods of observation. Documentary sources included holdings in the Western Cape Archives, the National Library, UCT's Jagger Library, and Stellenbosch University's research collection.

6 Pratt suggests this as an alternative to the notion of 'frontier' which foregrounds a European expansionist perspective. I am particularly indebted to Brenda Cooper for reminding me of this. See also Tuck and Yang (2018).

References

Christie, P. (2013). Space, place, and social justice: Developing a rhythmanalysis of education in South Africa. *Qualitative Inquiry, 19*(10), 775–785.

Christie, P. & Collins, C. (1982). Bantu education: Apartheid ideology or labour reproduction. *Comparative Education, 18*(1), 59–75; reprinted in Kallaway (1984).

Christie, P. & McKinney, C. (2017). Decoloniality and 'Model C' schools: Ethos, language and the protests of 2016. *Education as Change, 21*(3), 160–180.

Dubow, S. (1989). *Racial segregation and the origins of apartheid in South Africa.* London: Oxford University Press.

Evans, L. (2012). South Africa's Bantustans and the dynamics of 'decolonisation': Reflections on writing histories of the homelands. *South African Historical Journal, 64*(1), 117–137.

Everatt, D. (1992). Alliance politics of a special type: The roots of the ANC/SACP Alliance, 1950–1954. *Journal of Southern African Studies, 18*, 19–39.

Fuller, B. (1991). *Growing-up modern: The Western state builds third-world schools.* New York; London: Routledge.

Hamilton, C., Mbenga, B., & Ross, R. (Eds.). (2012). *The Cambridge history of South Africa, volume 1: From early times to 1885.* Cambridge: Cambridge University Press.

Hopkins, A. G. (2008). Rethinking decolonization. *Past & Present, 200*, 211–247.

Kallaway, P. (Ed.). (1984). *Apartheid and education: The education of black South Africans.* Johannesburg: Ravan Press.

Lefebvre, H. (1991). *The production of space.* Oxford: Blackwell.

Maldonado-Torres, N. (2007). On the coloniality of being: Contributions to the development of a concept. *Cultural Studies, 21*(2–3), 240–270.

Mamdani, M. (1996). *Citizen and subject: Contemporary Africa and the legacy of late colonialism.* Princeton, NJ: Princeton University Press.

Mignolo, W. (2007). Delinking: The rhetoric of modernity, the logic of coloniality and the grammar of de-coloniality. *Cultural Studies, 21*(2–3), 449–514.

Ngũgĩ wa Thiong'o. (2012). *Globalectics: Theory and the politics of knowing.* New York: Columbia University Press & Ebooks Corporation.

Posel, D. (1997). *The making of apartheid, 1948–1961: Conflict and compromise.* Oxford: Clarendon Press.

Pratt, M. L. (2007). *Imperial eyes travel writing and transculturation* (2nd ed.). New York: Routledge.

Ross, R., Mager, A. K., & Nasson, B. (2012) *The Cambridge history of South Africa, volume 2: 1885–1994.* Cambridge: Cambridge University Press.

Saunders, C. (2000). The transitions from apartheid to democracy in Namibia and South Africa in the context of decolonisation. *Journal of Colonialism and Colonial His*tory, *1*(1), 1–17.

Smith, L. T. (2012). *Decolonizing methodologies: Research and Indigenous peoples* (2nd ed.). London; Dunedin: Zed Books; Otago University Press.

Tuck, E., & Yang, K. (2018). *Indigenous and decolonizing studies in education.* New York: Routledge.

Wolpe, H. (1972). Capitalism and cheap labour power: From segregation to Apartheid. *Economy and Society, 1*(4), 425–456.

2 Schooling and inequality

Rhythms of sameness and difference

More or less in the middle of South Africa in the semi-arid Karoo, and away from the major highways that connect the north of the country with its ports to the south and the east, lies the small town of Carnarvon. History has left many traces in the town: in its wide and dusty streets, imposing church buildings, blockhouse on the hill from the Anglo-Boer War, racially segregated town planning, and separate 'location' for poor Coloured[1] people over the dusty riverbed just outside of the town to the north. Evident also are the markers of the rural economy that Carnarvon forms part of, with its obvious power differentials: expansive white-owned sheep farms for which the town serves as a hub (one sheep per six to ten hectares to be sustainable, I'm told), the groups of Coloured men on street corners waiting for casual farm work, and the gatherings of unemployed people whiling away their time outside the local pub. These markers point to historical inequalities that persist well beyond the official demise of apartheid, albeit in altered forms in the new dispensation. As a gesture to the future, a sign saying 'SKA in Africa' heralds the arrival of the world's largest radio telescope into the vicinity of the town. Considering these markers, what becomes immediately clear is that identities and meanings in daily life are not singular in a place like this. Intimations of possession and dispossession, and of 'making do' in this place, point to a complexity that seems to be in inverse proportion to the size of the town.

There are two schools standing on opposite ends of the town, each with its own distinctive apartheid architecture. On the northern end stands the primary school on the edge of the former Coloured township just outside the town, while on the southern end stands the secondary school in a suburban, former white area. The primary school was originally built in the apartheid days as a secondary school to serve the Coloured community. It has several banks of double-storey face-brick classrooms, linked by covered walkways and separated by strips of open area. Though space is abundant, the school itself has no particular facilities beyond a sandy sports field with goalposts and smudged athletics lines.[2] At break time, children play between and around the buildings on concrete or sandy school grounds. The secondary school has far grander double-story painted buildings that are spacious but clearly in need of maintenance. In former days, it was a 'combined' school for white students from Grades 1 to 12, but now some of its classrooms stand empty. The tennis courts and town swimming pool adjacent to the school are no longer maintained, and the remains of a shooting range speak to a time when white students won national prizes for the target skills they learned at school. In 2017, according to the Department of Basic Education, the primary school catered to 1,244 students with 52 teachers, and the secondary school catered to 447 students with 25 teachers.

There are no white students in either school, though white presence is evident in the town's shops and businesses as well as in the ownership of local farms. Almost all the students are Afrikaans-speaking and Coloured. I learned that some white children are registered for homeschooling and gather together in a 'support centre' in an unmarked house in the suburbs, while other white children attend boarding schools elsewhere. There are no institutions for post-school education or training to be seen. This means that to pursue further education, young people must leave town, which, in turn, means that mobility and its financing are key factors in education futures.

Official records of performance from the education departments show that the results achieved by the town's two schools are generally poor: there is high failure and repetition rate in the primary school, and a mere handful of students in the secondary school achieve the results required for university entrance. In terms of the official poverty classification on a quintile basis described later in this chapter, the primary school is classified in quintile one (most poor), while the secondary school is in quintile four (much less poor) – yet both schools clearly serve the same community. Census statistics suggest that this is a community of relatively high poverty.

In the town of Carnarvon, inequalities and rhythms of activity are easily visible in social spaces, and they point to the imprint of apartheid and colonialism before that. However, it is not straightforward to explain how these inequalities are produced, maintained, or changed decades after the end of apartheid; how schooling operates in relation to these broader inequalities; or the particular effects of place on schooling. To move beyond description requires fuller exploration of the social and material relations that make up spatial practices and their historical patterns, accompanied by an analysis of how schooling folds into them. It also requires an understanding of local place and its schools in relation to the broader national and global processes which impact upon them.

This chapter sets out the conceptual and theoretical premises used in this study within my chosen framework of de/coloniality, as outlined in Chapter 1. Since my intention is to use these theories in an embedded way, rather than to foreground them, readers may choose to pass over the next section in order to reach the main narratives and analyses of later chapters.

Theory's place

In setting out the theoretical premises of this study, I consider the following in sequence: first, 'sameness and difference' as the template of schooling and how school funding differs; second, the different scales of operation of a 'schooling system' that need to be considered while interpreting the outcomes of schooling; next, the importance of place and the relationships of influence between place and schooling; and finally, aspects of Lefebvre's (1991) analytic on the social production of space.

Schools: same same but different

Early each morning during the school term, groups of high school students make their way through the town of Carnarvon to the school on the suburban southern side. Some walk singly, and some walk in groups. Some stride purposefully, and some amble at a lazy pace. All are dressed in standard uniform: grey pants or maroon skirts and white shirts, maroon-coloured blazers or tops, and standard black shoes. In the other direction, children walk to the primary school, located across the dry riverbed at the edge of the township. Many walk from the northern and western sides of the town along the road and to the school. Some walk along the criss-cross paths over the vacant land and riverbed that separate the town from the township. Many walk from within the township itself. Their uniforms are similar to those of the high school, except their colour is green. As the school sirens sound their first calls of the day, the last stragglers scramble in. The directional movements of young people walking to school lend a sense of focus and purpose to the day in the town. For school-goers, this rhythm will continue in timetabled form until the final sirens ring in the early afternoon.

This rhythm is the masterplan for schools across the country, and the multiple variations it generates provide the experiences and outcomes of schooling – experiences that are lived in a variety of different conditions but generate fairly predictable outcomes when viewed at scale. Broad social patterns of class, race, region, former education departments, and gender are clearly visible in education statistics, though they are actually produced through diverse experiences in local places such as this.

To the urban eye, the stretched out scale of the vast and sparse Karoo expanses stands in strong contrast to the pushed-in scale and density of a city. The image of a concertina comes to mind – the movements of stretching out and pushing in are in fact different parts of the same instrument. If schooling is the instrument, its official script seems to operate with little or no regard for differences of scale or place. The rhythms of school days, terms, and holidays are marked out in repetitive, recognisable patterns, while the standard curriculum and national assessments provide an intended common pacing for teaching and learning across the country. Yet, for all the familiarity of form and rituals in schooling – uniforms, school bells, teachers, students, classroom spaces – the differences between schools are also striking. Step into any school, and its singularity soon becomes apparent. There are clear indicators of school culture: how the school deals with its litter; how achievement is displayed in trophies and pictures in the foyer or principal's office; the tone of the exchanges between the principal, teachers, and students; the pace of movement between classes; and the degree of effort taken to turn classrooms into spaces of learning. Indicators such as these point to a multiplicity of daily practices that escapes the singular logic of policy. Bourdieu's (1967) notion of the master patterns of culture (which he calls '*topoi*') that generate multiple variations of invention, much as a musical scale generates improvisation, may be usefully applied to schooling, whose master patterns generate a great variety of different forms on a single template. Each school has its own tonality in the cacophony of sounds on the same scale.

In South Africa, resource differentials between schools are particularly evident – an enduring legacy of the country's colonial and apartheid past. Over the years, successive apartheid governments had built and expanded a racially divided schooling system with unequal levels of provision, unequal expectations, and unequal outcomes – a master pattern of racial and spatial differentiation. Schooling was administered by racially divided education departments and subsequently by separate Bantustan administrations – a proliferation of spatial-racial departments across the country, all funded at different levels. Inequalities pervaded every aspect of schooling and its quality, from the material provision of schools and classrooms to the qualifications of teachers, from the adequacy of learning materials to the supply of electricity and toilets. Schooling was only compulsory for white children, and rural areas with high populations were particularly poorly provisioned – with differences also between Bantustans.[3] By 1994, the apartheid government was spending four times more on education for a white child than it was for a black child, and in the heyday of apartheid this was as much as twelve times (Figure 2.1).

To correct these distortions has been a major challenge for successive post-apartheid governments. More than two decades after the end of apartheid, inequalities continue to bedevil the schooling system, though the master pattern has been changed. The new master pattern highlights equality

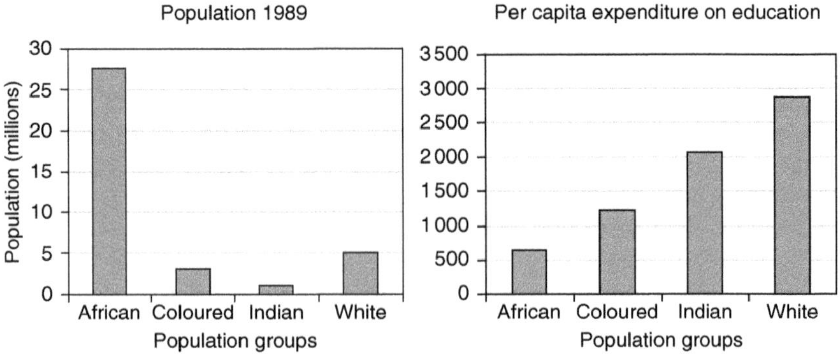

Figure 2.1 Per capita expenditure on different population groups, 1989.
Source: Own calculations.

as sameness in a single system – but in terms of funding, sameness applies only to the basic funding provided for all public schools. Added to this is a provision for schools to charge fees, and a small amount of non-recurrent funding is allocated to schools on an equity basis. In other words, the funding structure for public schooling allows for a difference in the resources made available to public schools, alongside the sameness of basic provision. Additional resources mean that fee-charging schools have more teachers, greater subject choice, and more provision for sporting and cultural activities in which to engage students beyond the formal curriculum (I explore these points in later chapters). Apartheid legacy black schools typically have larger class sizes, limited curriculum options, and no sporting facilities. In fact, 80% of their students are supplied with meals by the national school nutrition programme (StatsSA, 2018).

Though the multiple racial departments have been scrapped and replaced by nine provincial departments, this different configuration does not change the physical location of the schools or the material base on which improvements can be made. Provinces with large rural areas incorporating former Bantustans are home to the majority of the country's poor schools in terms of provisioning and official poverty status – as well as to poor learning outcomes. Provinces that incorporate multiple former education departments have had particular difficulties in administering their schools and improving their performance. An added complication is that funding is provided by the national government to provinces, who then have the responsibility of allocating the funds to schools – and this does not always happen as envisaged by the national government. The result is that even though there have been notable improvements in the schooling system, the overall quality of schooling is extremely poor and still reflective of historical racial and regional inequalities.

Different funding for different quintiles

The extent to which current inequalities within the system mirror the apartheid past is evident in a closer consideration of the quintile system for equity funding. As mentioned, schools are classified into quintiles from poorest (quintile 1) to least poor (quintile 5) for equity funding for non-recurrent expenses. Quintiles were initially drawn up on a provincial basis, but this was changed to a national basis after 2007 in acknowledgement that poverty is not proportionally spread over all provinces. The uneven spread of poverty Quintiles across provinces is clearly evident in the Department of Education's figures for equity funding for 2018 (Table 2.1). These figures show that the bulk of the poor schools (Quintiles 1 and 2) are in Limpopo and the Eastern Cape – provinces that have high rural populations and include former 'independent' Bantustans. In other words, the poorest schools under apartheid remain the poorest schools in the post-apartheid system, though they are configured differently: this time in terms of poverty level rather

Table 2.1 National poverty distribution table

National Poverty Distribution Table						
	Quintiles %					*Total*
	1 (Poorest)	*2*	*3*	*4*	*5 (Least Poor)*	
Eastern Cape	27.3	24.7	19.6	17	11.4	100%
Free State	20.5	20.9	22.4	20.8	15.4	100%
Gauteng	14.1	14.7	17.0	21.9	31.4	100%
KwaZuluNatal	22.1	23.2	20.2	18.7	15.8	100%
Limpopo	28.2	24.6	24.2	14.9	8	100%
Mpumalanga	23.1	24.1	21.5	17.7	13.5	100%
Northern Cape	21.5	19.3	20.7	21.4	17.1	100%
North West	25.6	22.3	20.7	17.6	13.7	100%
Western Cape	8.6	13.3	18.4	28	31.7	100%
South Africa	20	20	20	20	20	100%

Source: DBE (2018).

than race and Bantustan. The Gauteng and Western Cape Provinces have the highest concentrations of quintile 5 schools – significantly more so than any other province – and these are the most urbanised provinces, with no former Bantustans and the highest numbers of former white schools. The Northern Cape Province (where Carnarvon's schools are located) has the most even quintile spread, and it is the province with the smallest population percentage (2%) and the largest geographical spread (30% of the country).

Taking this analysis a step further: the government has progressively increased the number of schools that are classified as 'fee-free' so that quintiles 1–3 and some of quintile 4 are not able to charge fees – amounting to 71% of the country's schools (StatsSA, 2018).[4] What this indicates is that a very high majority of schools are in communities estimated by the government to be too poor to pay fees. In 2019, the government's equity funding formula allocated funding in addition to its basic 'norms and standards' funding as follows: R1,390 per student to schools in quintiles 1–3, R697 per student to quintile 4 schools, and R241 per student to quintile 5 schools (DBE, 2018). Census estimates are that average annual fees paid in 2017 were R9,000 per child (StatsSA, 2018), though some public schools charge fees of nearly that per month. What this means is that the 'poorest' state schools have considerably fewer resources with which to perform alongside the 'least poor' state schools on the template of sameness. An insight into the extent of poverty in poor schools is the finding by Hall & Giesen (2008–2009, p. 40) that the revenue of 92% of low quintile schools in Limpopo increased by an average of R60,000 per year when they were declared no-fee schools since they had been unable to raise fees from poor communities prior to this injection of funding. Whether this level of funding – intended for non-personnel spending – is sufficient for a school to function effectively is another matter.

Providing the same basic funding for all schools and adding an equity supplement may be regarded as a pro-poor measure, but it is limited in its capacity to build equity, particularly since amounts are relatively small and intended for non-personnel funding. Importantly, there is no provision for redressing the historical resource inequalities between schools. These are particularly stark when former white schools are compared with their counterparts in rural and former Bantustan schools. In 2018, only 59% of schools met the required national minimum standards for physical infrastructure, 30% had no running water, 20% were without adequate sanitation, and 48% had no library access (StatsSA, 2018). Withholding funds from the 'least poor' quintiles 4 and 5 does not release funding for the backlogs to be remedied. On the contrary, it means that resource differences are likely to increase.

It is often remarked that the post-apartheid system has moved on from race to class in its differentiation, and this is certainly reflected in the profiles of public schools that are able to charge fees. However, this is not the full picture, given that the majority of the country's schools are much the same as they were in the past. To the extent that the education policy suite suits the profile of quintile 5 schools – schools that succeed in the system – it is not likely to fit the majority of the country's schools, which are poor, rural, and black-only.

In looking at the different starting points of schools that face the demands of the same academic curriculum, researchers Teese and Polesel (2003) make a valuable distinction between 'fortified' and 'exposed' schools. 'Fortified schools' are schools where parents of high economic status 'pool their resources' to improve the learning opportunities available to their children: they pay for better qualified and more experienced teachers, smaller class sizes, well-stocked libraries and electronic resources, extramural activities, and so on. Their pooled resources strengthen their schools' capacities to meet the demands of the academic curriculum in order to improve their chances of successful outcomes. By contrast, 'exposed schools' have no additional resource advantages to meet the demands that the curriculum makes of them. They have high concentrations of poor students, and their 'pooled resources' include multiple disadvantages – poor language skills, fragmented family lives, limited leisure opportunities, and poverty in general. For these schools, there is a big gap between 'what the academic curriculum assumes about students and who students really are' (p. 123). These are the schools that generally perform poorly against the demands of the academic curriculum.

Same curriculum, different outcomes

A closer look at learning outcomes in South African schools – test results and drop-out rates – shows that the apartheid template still holds valence. Ample research shows deep and persistent patterns of inequality that are linked to place and former apartheid education departments as well as class and race (and gender in varying ways).[5] There is general agreement in describing these patterns, though there are differences in explaining them. Stephen Taylor (2011), economics researcher and member of the Department of Basic Education, provides a clear summary of these research findings:

> The substantial increase in resources invested in the historically disadvantaged parts of the school system has unfortunately not produced a commensurate improvement in education quality. This is clearly evident in the test scores of South African students in numerous surveys of educational achievement that have been carried out in recent years. These surveys have unequivocally shown that the overall level of achievement amongst South African children is extremely low. When the data allows for a disaggregation of schools according to the historically different systems a massive disparity is clear. Consequently numerous authors have now described the distribution of educational achievement in South Africa as bimodal .... By this it is meant that the overall distribution in fact conceals two separate distributions corresponding to two very differently performing parts of the South African school system. Fleisch (2008) maintains that there are effectively two education systems within one in South Africa. The difference between the two systems is rooted in the historically separate administration of education for each race group. The majority of South Africa's students (80% – 85%) are located in the historically disadvantaged system and demonstrate very low proficiency in reading, writing and numeracy. The second system produces educational achievement that is closer to what would be expected in the developed world. This system serves mainly white and Indian children, and increasingly black and coloured middle class children. The vast majority of university entrants are produced by this latter system. Van der Berg (2008, p. 145) describes these two groups of schools as operating under 'separate data generating processes'. It is therefore important to be sensitive to this underlying structural aspect when analysing educational achievement data for South Africa.
>
> (p. 4)

Results from international tests also provide a particularly sobering reflection of overall system performance.[6] In the 2006 Program for International Student Assessment (PIRLS) test of reading literacy, South Africa's mean reading score was the lowest of the 40 participating countries. In the 2003 Trends in International Mathematics and Science Study (TIMSS) tests, South Africa's mean scores in mathematics and science were the lowest of the 50 participating countries. In the 2017 TIMSS test, South Africa moved from last to second-to-last of the 41 participating countries, prompting the national education department to make the absurd claim in its press release that 'South Africa has been the fastest improving education system in TIMSS between 2002 and 2015'. This claim brings into question the base from which improvement is measured, to say nothing of the pace of change.

In short, statistical analyses show beyond any doubt that the opportunities that schools afford to students depend on who the students are, where they go to school, and what schools they attend. Even although all public schools in South Africa are equal under the law, the law also enables them to be different. The meme 'same same but different', commonly used in South-East Asia, captures this ambiguity well, particularly since it usually signals an appearance of similarity but a difference in quality. This meme reflects a shared tacit agreement that all schools can be viewed as the same, accompanied by a shared tacit acknowledgement that there are structural and substantive differences between them.

Schooling systems on different scales: global, national, provincial, local

A premise of this study is that the systemic provision of schooling operates at different levels or on different scales, and understanding these is important both for studying policy and for achieving change. On a national scale, the policies that regulate the system as a whole are guided by macro-level logics: for example, in decisions about resources and procedures for allocating them, and in regulations to direct who should teach what to whom. Policies are often compromises that stitch over different views and values, and they are translated – often poorly – into laws and regulations. Today, when national policies are influenced by global policy trends and policy-borrowing is common, there are homogenising effects which tend to gloss over local contexts – a point I return to later. Bureaucracies responsible for national education policy development and implementation are multilayered and complex, and national policies must weave their way through provincial (or sub-national) departments to reach the smallest units of the system – schools and classrooms.

It is on this smallest of scales that the central purpose of the system as a whole is given expression. On this smallest scale, teaching and learning in classrooms are governed by micro-level logics and dynamics quite different from the macro-logics and policies of the system, and necessarily so. This means that one of the abiding challenges of education policy is to ensure that what is intended on one scale is enacted on another. Another abiding challenge is to ensure a good-enough match between national policy visions and the actual conditions of schools and classrooms since it is this smallest unit that ultimately produces the successes or failures of an education system.

It is important for education researchers to be alert to the multiple activities, dynamics, and relationships that operate in different ways across policies as they are enacted through nationwide bureaucracies to reach specific schools and classrooms. Policies, including equity policies, are seldom enacted as envisaged, and 'things are often not what they seem'. Specific social contexts have very significant effects on the smallest units of the system (schools and classrooms), and this, too, needs to be accounted for in designing research on schooling.

Having said that, research in sociology of education over decades has shown conclusively that schools are so imbricated in broader social contexts that they have limited ability to change these. So predictable are the patterns of performance in schooling outcomes that it seems disingenuous not to consider them when policies are designed, or results are compared.

In judging the performance of schools, it is common to base comparisons on political units, such as nation states or provinces, without specifically considering the local circumstances of the schools. This can be illustrated on national as well as global scales. A simple illustration on the national scale is the annual comparison of the results of the National Senior Certificate (NSC) across South Africa's provinces and the ritual 'praise and shame' in local media. (The same holds to a lesser extent with Annual National Assessments (ANAs) in Grades 3, 6, and 9 when they take place.) Results are presented as if they are simply the product of provincial education departments, yet in fact they are also the product of particular schools and – as mentioned – their socio-economic contexts as well as the legacy department they were part of under apartheid (see Spaull, 2013; Christie, Butler & Potterton, 2007). When national and provincial performance data is presented, the effects of class, race, gender, and language are easily discernible, while the particularities of schools disappear. The aim of this study is to restore the local level to a position of visibility. From the vantage point of the local, different insights emerge on schooling, continuity, and change.

On a global scale, the fashion of international comparative tests such as PISA, hosted by the Organization for Economic and Cultural Development (OECD), similarly displaces the local from consideration.[7] Instead, international comparative tests may give the impression that countries' performance scores can be straightforwardly compared, implying that the experiences of schooling are homogenous rather than context-related in the different countries participating in the tests. Like national tests, these global comparative tests may gloss over the influence of local conditions, and as they do so, they reinforce the view that contexts are neutral.

Yet there is also significant research, ironically even within the PISA testing process itself, that points to the overriding effects of socio-economic, linguistic, and cultural differences in producing the unequal outcomes of schooling. For example, the 2018 website summary of the OECD (2018) report, *Equity in Education*, states unambiguously that:

> The report finds a strong link between a school's socio-economic profile and a student's performance: students who attend more socio-economically advantaged schools perform better in PISA. Yet, on average across OECD countries, 48% of disadvantaged students attended

> disadvantaged schools in 2015 and there has been no significant change in segregation levels in most countries over the past decade.
>
> On average across OECD countries, disadvantaged students attending advantaged schools score 78 points higher than those attending disadvantaged schools, equivalent to more than two and a half years of schooling.

Though there are strong international and comparative studies that acknowledge the effects of socio-economic factors on school and student performance, few consider what redress might entail. There are warnings in more critical studies, though, against assuming that there are 'lessons' that are easily transferable from one country to another – a point well argued, for example, in the articles in the special edition on PISA in the *Teachers College Record.* In their introduction to this issue, Perry and Ercikan (2015) usefully caution against 'inappropriate interpretations' (p. 3) of PISA results, such as the assumption that the practices of some countries may be used 'to identify strategies to improve learning in other countries'. In terms of considering what change might entail, Willms's (2006) study, *Learning Divides*, offers useful pointers. While looking at the inequalities between and within selected countries, together with their test achievement scores, Willms suggests that a mix of differentiated strategies may in fact be needed 'to raise and level the learning bar' in different contexts, and he provides examples of what these might entail.

Place and schools

Places are the particular contexts within which schools take their specific forms – their own tonality. Places are where people live their daily, embodied lives, shaped by both their geographic location and their histories. Places are mapped and represented in surveys and plans that illustrate power relationships in material, representational, and symbolic ways. They are also important sources of identity and meaning, of feelings and imagination, and of artistic and symbolic expression. Extending this point, I argue that places also influence the actions, intentions, and experiences of their schools. Further still, I argue that the histories of places and their specific social relations have enduring significance, and that histories of schooling form a background or palimpsest often still visible in present arrangements.

The focus of this study is on shared rather than individual perceptions, social imaginary rather than personal consciousness. There are inevitably individuals with aspirations and ambitions that transcend their particular social contexts, and while it is important to acknowledge this, such individuals are not the focus of this study. My concern is with shared social expectations of schooling and its opportunities, which go beyond a simple aggregation of individual experiences. I focus on the jointly held sense of the structure of opportunities for people – in this case, people living in a particular place, based on local circumstances, making arrangements for schooling with the resources at their disposal.

In looking at the opportunities offered by schools in specific places, Gibson's (1979) concept of affordances is helpful. Writing in the context of social psychology, he proposed a mutuality of relationships between animals (including humans) and their environments. He wrote:

> The '*affordances*' of the environment are what it *offers* the animal, what it *provides* or *furnishes*, either for good or ill. The verb to *afford* is found in the dictionary, the noun *affordance* is not. I have made it up. I mean by it something that refers to both the environment and the animal in a way that no existing term does. It implies the complementarity of the environment and the animal.
>
> (p. 127)

Applying this notion of complementarity to the relationships between schools and their communities, 'opportunity' becomes a multi-sided notion.[8] It refers both to what schools actually provide and to what people perceive them to provide. Thus, it requires consideration of *both* the material practices and circumstances of schools, *and* the perceptions and imaginaries of what they provide. As used in this study, the notion of affordances highlights the nexus between what schools actually offer and the shared social perceptions of what they might offer in a particular place and time.

Affordances of schooling are influenced not only by what happens inside schools themselves but also by a range of contextual factors, such as their history, the structure of local economy, possibilities for civic participation, opportunities for out-of-school activities, and linkages to other places. National policies have material effects on schooling in local places, albeit in complex and differentiated ways, and national policies are themselves influenced by global trends. What becomes apparent in this expanded thinking is that there is a complex nexus of interacting relationships that influence the opportunities that schools offer and those that people perceive them to offer. It is this complex nexus that I refer to with the 'affordances of schooling'.

While emphasising place and affordances, it is important to recognise that 'place' does not determine the activities and outcomes of specific schools. My own work on 'resilient schools' (Christie, 2001) and 'schools that work' (Christie et al., 2007) has shown that there are schools that survive and thrive in the difficult contexts in which other schools fail. Around the corner from a 'dysfunctional school' will be a 'resilient school' that manages to succeed in the experiences it provides to students and staff, and in the outcomes it achieves. In short, places do not determine the functioning of schools and their outcomes, but these cannot be properly understood without considering places and their local dynamics. Schools as individual organisations need always to be understood as both similar to and different from each other – as in the meme 'same same but different'.

Schools in social space

To examine the different strands of practice that make up social space, Lefebvre's works on *The Production of Space* (1991) and *Rhythmanalysis* (2004) provide a guiding framework (though this is loosely used). Lefebvre insists that space should not be conceived of as an empty container that can be separated from its contents. Nor is it inert, like an object. Rather, social space is *produced* through activities and practices, grounded in geography, and shaped by history and symbolic meanings. Lefebvre identifies three interwoven sets of practices that make up social space: the recognisable movements, routines and activities of daily life (comings and goings which he terms 'perceived space'); the maps, plans, or policy formulations that are abstractions or representations of these activities (Lefebvre calls this 'conceived space'); and the lived, embodied experiences of people, infused with imaginary and symbolic elements (this is Lefebvre's 'lived space'). When taken together, this set of perceived-conceived-lived practices provides an overarching framework for analysing the different practices of social space. A strength of this unitary framework is that it enables different, interrelated practices to be taken together but examined in terms of their own rhythms and complexities, without collapsing them into each other.

A distinctive feature of Lefebvre's (1991) approach to the production of space is its sensitivity to movements and rhythms, and to patterns of repetition and difference in social practice. Thus he talks of space in terms of myriad flows, waves, and movements:

> The form of social space is encounter, assembly, simultaneity … [of] everything that there is *in space*, everything that is produced either by nature or by society, either through their co-operation or their conflicts. Everything: living beings, things, objects, works, signs and symbols.
>
> (p. 101, original emphasis)

In space-as-encounter, there are different rhythms of practices taking place simultaneously on different scales – global, national, regional, local. These interpenetrate, superimpose, collide, and clash. So, in the case of education policies, it is possible to trace global trends and influences on national policies as well as the interrelationships of national and provincial policies, and the rhythms of local practices. Lefebvre is insistent that local place continues to exist as a particular point, even though its practices may shift as a result of the polyrhythmic flows and movements on other levels.

Lefebvre's (1991) framework seeks to provide a unitary theory that enables the analysis of a range of different interpenetrating practices on different scales (global, national, local) without collapsing them or flattening them out. An important point he makes is that:

> The principle of the interpenetration and superimposition of social spaces has one very helpful result, for it means that each fragment of

> space [when] subjected to analysis masks not just one social relationship but a host of them, that analysis can potentially disclose.
>
> (p. 88)

The challenge, then, is to investigate schools in particular places and times as 'fragments of space', and to analyse the different social relationships and rhythms, movements, and practices on the different scales that play across them in ways that are not always fully predictable.

Viewed in terms of this framework, it is clear that schools as social spaces are replete with social practices and regularities of many sorts.[9] There are familiar daily routines with different rhythms of teaching and learning taking place in demarcated spaces. The recognisable architectural forms of schools are tailored to the standard purpose of single adults conveying socially valued knowledge to groups of students in a standard classroom format. In South Africa, the architecture of apartheid is evident in schools designed and constructed in the past for the different racial departments. Schools are regulated by policies – representations of the intended activities of formal teaching and learning. Students' lived experiences are intentionally disciplined, with bodily formation taking place in ritualised repetitions as people sit at desks and take part in shared activities, such as assemblies, sports, choirs, and so on.[10]

Yet for all their similarities of form, schools are also constituted in patterns of difference. Different relationships to school knowledge are evident in different schools; different bodily gestures and even handwriting speak to different 'dressage'. Place-specific activities (such as festivals and exhibitions – or even their absence) engage schools with their local environments in specific ways, and different place-based social and economic activities impact upon the affordances of schooling. In Lefebvre's terms, all of these practices are part of 'the encounter of everything' in the social space of the school or the school in its social space. Education policies on different scales – national, provincial, and district – act across the practices of schools in different ways, and these, in turn, are informed by global representations of what constitutes 'best practice' and satisfactory performance.

Taking Lefebvre at his word that 'fragments of space' may be analysed to show a host of social relationships, I approached my research on the schools in Carnarvon with a view to discerning the rhythms of schooling in the 'encounter of everything' in this particular small town in a semi-arid and marginal place, with its own history and symbolic meanings, and identifiable practices of daily life.

Carnarvon is a town in which colonisation and apartheid have shaped the political economy and social relations of space, and their imprint is still clearly visible, more than twenty years after apartheid's formal demise. The town and its schools enable close consideration of the puzzling question facing post-apartheid South Africa: why have inequalities in schooling been so hard to change, in spite of policies designed to shift schooling away from the divisions of apartheid towards a more equitable system of education for all? Against the general studies of history and policy analysis in education, this study of Carnarvon looks at this puzzling question.

This book traces the narrative of the schools in the town of Carnarvon as a vantage point for closer inspection in larger studies of policy analysis and history. An in-depth study like this enables us to magnify the logics of policy in a particular context, it enables the palimpsest of colonial schooling to be more clearly seen, and it provides insights into how and why policy enactments do not match their stated intentions. While the study of a single place over time does not provide a basis for generalisation, it nonetheless provides insights that are not easily available in general histories and comparative statistical analyses. Detailed engagement at this level also provides a valuable counterpoint to broader theories of change – including postcolonialism and decoloniality – as they come to grips with what may be involved in shifting beyond the inequalities and exclusions of colonisation and its continuation in forms of coloniality.

Possession and dispossession in place

Bringing these theoretical strands together to craft an account of schooling and place, this study analyses two 'focal point' moments: the early establishment of the town of Carnarvon and its schools in the period before and after 1860 as part of colonial expansion, and the contestation over the schools from the 1990s to the present as part of the process of dismantling apartheid.[11]

The first of these periods shows the establishment of the social and economic relationships of settler colonialism and the forms of schooling at the time. The second focuses on the changes after apartheid ended – in political arrangements and in schooling that enabled the intersectional inequalities of colonialism to persist into relationships of coloniality. In discussing this second period, I consider the SKA as 'big science' meeting development in this local place. For both periods, I provide a brief account of the distinctive social relationships formed in Carnarvon, showing how particular themes of possession/dispossession have reverberated across time. And for both periods, I link these broader social arrangements to the provision of schooling and its goals. To link the two periods, I provide a chapter summing up developments between them.[12] The intention of the study is to analyse the town and its schools as 'fragments of space' in two time periods, to show how inequalities of schooling are formed and enacted within the specificities of place and time, and to show how historical relationships established in colonial times linger on. In the conclusion to the book, I explore possibilities for shifting these deep, structural inequalities within the spaces opened up by decoloniality theorists and the challenge they pose in order to shift the form of the conversation as well as its content.

During fieldwork in the town, an intriguing mantra was repeated by a number of white interviewees: '*First they took our municipality, then they took our school, and now they want our church*'. The mantra of zero-sum patterns of 'us and them' provides a thread that I sustain through the chapters that follow, as I explore issues of decolonising schooling from the vantage point of this place.

Notes

1 Unfortunately it is not possible to talk of apartheid and its consequences without reference to its racial classification system (as mentioned at the start of this book). 'Coloured' is the colonial and apartheid term for people classified as mixed race, and I use it with this historical connotation, albeit with reluctance.

2 The Department of Arts and Culture built a basketball facility next to the sports field at the end of 2015, but on my research visit in 2016 it was not in use. The sports facility across the road had not yet been completed; it had seating and (presumably) provision for ablutions.

3 Very little research has been carried out on schooling in Bantustans – with the notable exception of unpublished research reports by Heather Jacklin and Johann Graaff.

4 StatsSA (2018) estimates that 67.2% of students attended schools where no fees were levied, and a further 6% paid fees of less that R100 – 'tantamount to fee free schooling' (p. 38). The breakdown by province is: Limpopo 92%, Eastern Cape 78%, Western Cape 47%, Gauteng 51%. Another indicator of poverty is school feeding, provided to 91% of students in Limpopo, 90% in Eastern Cape, 88% in Mpumalanga, 54% in Western Cape, and 55% in Gauteng.

5 Extensive research has been done on this by the Programme to Support Pro-Poor Policy Development, a partnership between the Presidency and the European Union, and by researchers from the Research on Socio-Economic Policy (RESEP) at Stellenbosch University, led by Servaas van der Berg. Useful papers by Martin Gustafsson, Debra Shepherd, Nic Spaull, and Steven Taylor (among others) appear in this research series and the Stellenbosch Economic Working Paper series.

6 International tests include Progress in International Reading Literacy Study (PIRLS) and Trends in International Mathematics and Science Study (TIMSS). For a summary of South Africa's performance on international tests in 2015, see Africa Check (2015) and the OECD's (2019) statistical summary.

7 PISA tests are conducted triennially to assess the performance of 15-year-olds from over 90 countries in mathematics, science, and language. South Africa has an extremely poor record of performance on PISA. Though PISA tests do collect data related to context, reports on this are seldom referred to in the overall presentation of scores.

8 I use the term 'community' cautiously, recognizing, after Bauman, that communities are not without their tensions and trade-offs. 'Neighbourhood' is a potentially more neutral term, referring as it does to geographical contiguity. Yet I would suggest that schools do in fact operate in communities, with all their contradictions, rather than in neighborhoods, with their apparent neutrality.

9 In my earlier work on dysfunctional schools (Christie, 1998), what was strikingly apparent to me was how dependent schools are on the demarcation of time and space. Even where there are no buildings, such as schools under trees, those entering the symbolic space of a classroom will gesture a knock.

10 Lefebvre (2004) describes activities of bodily formation such as these as 'dressage', and they point to the formation of 'habitus' and 'cultural capital' in Bourdieu's terms.

11 The design of the study has similarities to the 'laminated analysis' of critical realism. It also has similarities to the 'transversal comparative case study' of Bartlett and Vavrus (2017), with the two time periods counting as cases in a study of place across time. However, its design does not follow these more structured approaches. Its underpinning shape is informed by a Lefebvrean approach – to trace rhythms of practice in time and space – and the theoretical framing in sociology of education and de/coloniality outlined earlier.

12 It needs to be said at the outset that the selection of two time periods for analysis sets definite limits for this study, as does the decision to narrow the analysis of schooling to issues of provision and social meaning. These limits are necessary for pragmatic rather than theoretical reasons; without such limits, the task would simply be too vast to complete in a single study. The study does not attempt to be comprehensive in scope, nor does it claim to show all of the movements and flows that have shaped the town and its schools over time.

References

Africa Check. (2015). Assessing South Africa's schooling system. https://africa-check.org/factsheets/guide-assessing-south-africas-schooling-system/ Accessed 19 November 2019.

Bartlett, L. & Vavrus, F. K. (2017). *Rethinking case study research: A comparative approach*. New York: Routledge.

Bourdieu, P. (1967). Systems of education and systems of thought. *International Social Science Journal*, *19*(3), 338–352.

Christie, P. (1998). Schools as (dis)organisations: The 'breakdown of the culture of learning and teaching' in South African schools. *Cambridge Journal of Education*, *28*(3), 283–300.

Christie, P. (2001). Improving school quality in South Africa: A study of schools that have succeeded against the odds. *Journal of Education*, *26*, 40–46.

Christie, P. (2008). *Opening the doors of learning: Changing schools in South Africa*. Johannesburg: Heinemann.

Christie, P. (2016). Educational change in post-conflict contexts: Reflections on the South African experience 20 years later. *Globalisation, Societies and Education*, *14*(3), 434–446.

Christie, P. & Collins, C. (1982). Bantu education: Apartheid ideology or labour reproduction. *Comparative Education*, 18(1), 59–75.

Christie, P., Butler, D., & Potterton, M. (2007). *Report of the ministerial committee on schools that work*. Republic of South Africa: Department of Education.

Department of Basic Education (DBE). (2018). *Amended norms and standards for schools funding*. Pretoria: Government Gazette, 2 February 2018, no. 75.

Gibson, J. J. (1979). The theory of affordances. *The ecological approach to visual perception* (pp. 127–143). Hillside, NJ: Laurence Erlbaum.

Hall, K. & Giesen, S. (2008–2009). Addressing quality through school fees and school funding. In Children's Institute (Ed.), *South African Child Gauge* (pp. 35–40). Cape Town: UCT.

Lefebvre, H. (1991). *The production of space* (D. Nicholson- Smith, Trans.). Oxford: Blackwell.

Lefebvre, H. (2004). *Rhythmanalysis: Space, time and every-day life* (S. Elden & G. Moore, Trans.). London: Continuum.
OECD. (2018). *Equity in education*. www.oecd.org/education/equity-in-education-9789264073234-en.htm Accessed 12 November 2019.
OECD. (2019). *South Africa*. http://gpseducation.oecd.org/CountryProfile?primaryCountry=ZAF&treshold=10&topic=EO Accessed 19 November 2019.
Perry, N. & Ercikan, K. (2015). Moving beyond country rankings in international assessments: The case of PISA. *Teachers College Record*, *117*(1), 1–10.
Spaull, N. (2013). Poverty and privilege: Primary school inequality in South Africa. *International Journal of Educational Development*, *33*, 436–447.
StatsSA. (2018). *General household survey 2017: Focus on schooling*. ISBN: 978-1-4315-3269-8.
Taylor, S. (2011). Uncovering indictors of effective school management in South Africa using the national school effectiveness study. Working Paper, University of Stellenbosch Department of Economics and Bureau for Economic Research.
Teese, R. & Polesel, J. (2003). *Undemocratic schooling: Equity and quality in mass secondary education in Australia*. Carlton: Melbourne University Press.
Willms, J. D. (2006). *Learning divides: Ten policy questions about the performance and equity of schools and schooling systems*. Montreal: UNESCO Institute for Statistics.

3 Colonialism, possession, and dispossession

The Karoo and its people

Beyond the coastal fold mountain ranges in the south, south-east, and south-west of South Africa rises the great escarpment of the Karoo Supergroup. In its eastern stretches, it forms the Drakensberg mountains, while its west and southern stretches are marked less dramatically by the flat-topped dolerite hills characteristic of the Great Karoo. Coastal mountains catch the rains, leaving the hinterland with little precipitation. In the western winter rainfall area, the interior is particularly dry, and vegetation is sparse. To the east, the summer rainfall pattern yields more precipitation to the interior plain, but even so, rainfall is irregular and unreliable, and drought is frequent. This arid and semi-arid hinterland has a distinctive landscape and biome, with plains of sparse and scrubby bush, interrupted by tumblings of dark stones patinated by extremes of temperature, and framed by distant blue flat-topped hills and vast skies. Before the advent of windmills to source underground water, this space could not sustain permanent pastoralism, though it could sustain movements of animals and people as the varying conditions of vegetation and water sources allowed.

The settlement of this particular place in processes of European colonisation is the subject of this chapter. My aim is to show how the distinctive, zero-sum patterns of settlement and displacement took place over time, shaping racial identity and class formation as the town of Carnarvon was established in the mid-1800s. Over this chapter and the next one, I develop the argument that the complex colonial relationships that developed here have had defining consequences for the town and its schools, reverberating into current times. Building on the narrative of this chapter, the next chapter shows in more detail the intimate relationships between schooling and place. I aim to show that space and place are not simply 'backdrops' to schooling, but are imbricated in its provision and meanings. And in Carnarvon, social and economic inequalities, imbricated in schooling, were part of the town's establishment.

The landforms of the Karoo, distinctive as they are, also preserve an ancient past. Research by palaeontologists shows a record of life in the Karoo stretching back more than 300 million years. Once a vast inland basin, the Karoo has fossils of the earliest life forms as well as of amphibians and early mammal-like

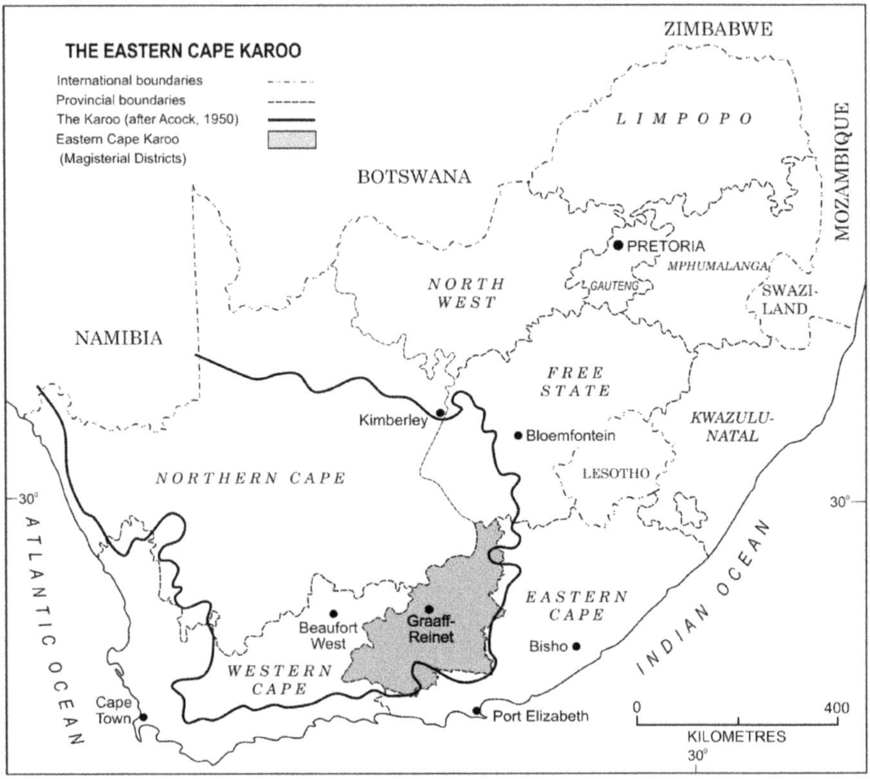

Map 3.1 Karoo.

Source: Reprinted from *Urban Forum*, 22, Nel, E., Taylor, B., Hill, T. & Atkinson, D. (2011). *Demographic and Economic Changes in Small Towns in South Africa's Karoo: Looking from the inside out*, p. 400, with permission from Springer.

reptiles. Its sequence of sandstone rocks has preserved 'a world-class assemblage of fossils, spanning a time-range from the Late Carboniferous to the Early Jurassic' (Rubidge & Hancox, 2002, p. 54). This assemblage shows the early evolution of dinosaurs, tortoises, and mammals. Not far from the present-day town of Carnarvon, the reassembled fossil of a Bradysaurus reptile lies preserved in a small wire-fenced enclosure on an out-of-the-way farm, Melton Wold – a species that roamed the earth more than 250 million years ago.

Archaeologists have found multiple material traces of the ancestors of hunter-gatherer people living in the Karoo area. Parkington, Morris, and Rusch (2008, p. 83) note that artefacts found in the Karoo – rock etchings, scatterings, tools, stones, and bones – are typical of Earlier and Middle Stone ages, with evidence dating from more than a million years ago. Late Stone Age artefacts are also plentiful. As well as hunter-gatherers in the southern and central areas of South Africa, archaeological evidence points to pottery-using, sheep-owning Khoekhoe moving into the area about 2,000 years ago.

Remarkably, in the case of the Karoo, archaeological records are supplemented by an archive of handwritten notebooks of San stories, drawings, genealogies, and other materials gathered by philologists Wilhelm Bleek and Lucy Lloyd in the 1870s from San people brought to prison in Cape Town under colonial law. The Bleek and Lloyd archive is an extraordinary collection that gives insights into San relationships with landscape, animals, skies, and each other. The archive has been brought to light by the creative art works of Pippa Skotnes (1996), the story telling of Manuel de Prada Samper (2012) and the archaeology of Deacon (1986), who retraced the sites of the Bleek and Lloyd informants, and others. Not only does archaeological research, supplemented by these archives, reveal the San 'stories that float from afar'; it also reveals the harsh conditions of life to which the San were reduced and their ultimate elimination as a distinctive group under the pressures of colonial expansion.

The place on which this study is centred, the Karoo town of Carnarvon, speaks to both its significance and its marginality, for the town is located on two springs at the edge of the Kareebergen hills, named Schietfontein and Harmsfontein in the 1700s. These springs, while invaluable for temporary sojourns, are located in an area that was too arid to sustain settled pastoralism at the time of early colonisation. The springs were visited by transient animals and San hunter gatherers, as well as by scatterings of Khoekhoe and traders and adventurers crossing the dry plains from water source to water source. The travel writings of Lichtenstein, Burchell, and others comment on the formidable climate of these arid lands, as well as the huge migrations of springbok and the variety of game that could sustain the livelihoods of San people.

Remarkably, in the same San lands that are traced in the Bleek and Lloyd archive, present day astronomers have seen a different set of possibilities. The stable rock platform, an intrinsically quiet environment, clear skies, and sparse human occupation afford the location of telescopes, including the construction of part of the world's largest radio telescope array (the Square Kilometre Array or SKA) near Carnarvon. The smaller South African Large Telescope (SALT) assemblage views the universe from a hill outside Sutherland, a few hundred kilometres to the south and west.

For current inhabitants of the Karoo, about a million people, of whom many were born there, the Karoo provides the source of livelihood and meaning of daily life. Femke Brandt (2014) describes current conditions in the regional economy as follows:

> Life in the Karoo has been determined by agricultural rhythms and all people's struggles have been related to contests over who owns and uses this semi-arid desert land. Access to land and natural resources has shaped relations and experiences in the countryside and still does for most people. Whether waiting along the roadside to be picked up for a casual job on a farm or while applying for a government house because commercial farmers gradually exit residential labour arrangements; one can only understand the heartbeat of the Karoo if the quaint

> tourist towns, the dusty townships and the surrounding farmlands are regarded as one whole.
>
> (n. p.)

In the rest of this chapter, I set out a brief account of the historical contestations over land and water in the mid-1800s in what was then the shifting northern frontier zone of the Cape Colony. I argue that colonial patterns reverberate into current arrangements not only in schooling but also in land ownership and institutions such as the municipality and the church in this town – even if as echoes of influence across time rather than strong determinants.

The northern frontier zone of the Cape Colony

Switching perspective from the central interior to the coastal plans of the south-western Cape where European colonial settlement began, the vista is very different. From this vantage point, Table Bay and its nearby plains are surrounded by belts of fold mountains running parallel to the southern and eastern coasts. The great escarpment and tablelands of the Karoo lie across the fold mountains and much further to the north. To the east, the coastal plains provide a stretch of arable land with good rainfall. On this land, lived a number of Khoekhoe groupings with cattle and sheep, as well as local San – groupings that did not survive the intrusion of white settlement (Christopher, 1976; Hamilton, Mbenga & Ross, 2012).

European settlement began with the establishment of a trading station at Table Bay by the Dutch East India Company in 1652, which was taken over by the British in 1806. (Both heritages, Dutch and British, are evident in the schooling that was established as the settlement expanded, as the next chapter will show.) As white settlement expanded eastwards from Cape Town, Dutch settler farmers established themselves on these arable lands as wheat and wine farmers, using the labour of indigent indigenous people and imported slaves. Around Cape Town, a complex settlement grew, with a mix of people and evident disparities of wealth and status. Here was the seat of colonial governmentality and its logics of practice.

Beyond the official boundaries of the Colony, *trekboere* (white Dutch/Afrikaner frontier farmers) spread out in sparse occupation of open grazing lands – the *trekvelden*. Whereas settler movement to the north was constrained by arid conditions, land to the east became increasingly attractive to white settler expansion until 1779, when it was met with a substantial Xhosa presence at the Fish River, resulting in a hundred years of war. The defeat of the Xhosa as a significant polity, the incorporation of their land into the Cape Colony, and their proletarianisation and impoverishment, provides one of the major frontier narratives of colonialism in South Africa. Another major narrative of colonialism is the Great Trek of 1836 – the movement of Dutch colonists with their households into the interior lands of South Africa away from British control – lands already occupied by different major African polities. These movements of settlers skirted around the arid interior of the Great Karoo.

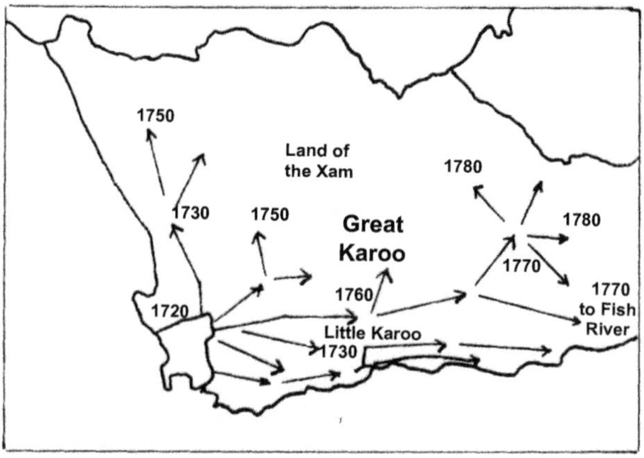

Map 3.2 Expansion of colonial graziers.
Source: After Christopher (1976).

Settler expansion to the area north of Cape town was constrained by its geography – fluctuating rainfall with periodic severe droughts, poor vegetative covering, and distance from the Cape settlement with limited means of communication. In this sparse area, colonial expansion had a violent impact, as shown in the work of Mohamed Adhikari (2011), Manuel de Prada Samper (2012), Nigel Penn (2005), and others. This work reveals the encounters of a mix of different people contesting the spaces in an unstable, shifting frontier zone: San hunter gatherers, Khoekhoe herders, *trekboere* (white Dutch/Afrikaner frontier farmers), people who identified as Basters,[1] Korana, Griqua, and even splinter groups of Xhosa – along with individual hunters, traders, fugitives from colonial rule, and also a handful of missionaries, as we shall see. Of prime value for people in this zone were access to water and grazing lands, and the ability to move between different sources of these according to season. Raiding and plunder of livestock was carried out across all groups, while cattle, ivory, and arms were valued items of trade. Penn (2005) aptly names this *The Forgotten Frontier* in his book of that title, reflecting on the focus of historical scholarship on the Eastern Cape as South Africa's defining frontier, to the virtual neglect of frontier patterns in this northern region.

This historical scholarship together with archaeological studies mentioned earlier shows the intensity of contestation over the scarce resources of water, grazing land and game as white *trekboere*, traders, and adventurers spilled over each border that was reluctantly extended by colonial authorities, first Dutch, then British. These settlers operated at the edge of, and beyond, the control of the law. The land that they spilled into, sparse as it was, was not 'empty': as mentioned earlier, for thousands of years it had been occupied by groups of San hunter gatherers, and Khoekhoe pastoralists also occupied suitable areas, often moving livestock in patterns of transhumance according to rainfall. These inhabitants were not passive in the face of colonial penetration into the area, as the records of attacks and counter-attacks of the late 1700s make clear. The San, in particular, were greatly feared as fierce hunters of the herded livestock that settlers brought into the land, displacing the animals they had hunted for food for thousands of years.

With colonial authorities reluctant to expend energy and resources in this frontier area, local government was administered by *landdrosts*/magistrates and *veldwagmeesters*/field-cornets who deployed a harsh commando system (a form of officially sanctioned citizen militia) to assert their power. The later years of the eighteenth century saw brutal, genocidal commandos against the San, who as Robert Ross (2012) notes, either fled beyond colonial control or were incorporated into colonial society 'at the lowest possible level, that of distrusted and exploited servant' (p. 200). Ross points out that the brutality and near-slavery with which farm labourers were treated meant that they 'regularly took up arms against their masters' (p. 203). Poorer white farmers were similarly displaced by capitalism as white commercial farmers – particularly wool farmers – took over much of the Karoo land by the end of the nineteenth century. It is this particular history of the possession and dispossession of the sparse Karoo resources – in different forms and phases – that reverberates in the present social arrangements of the smaller, post-apartheid Karoo towns.

The Xhosa settlement at Schietfontein

In the semi-arid area south of the Gariep/Orange River where different San groups roamed, there were by the late 1700s several small and moving pastoralist groups of Khoekhoe, Basters, Griqua, as well as white *trekboere*, often intermingling. Where settlement occurred, this was not in continuous waves but rather around sources of water. The dry backlands from these holdings were often vacant, with the land in between regarded as common pasturage.[2] Isolated from the south-western Cape settlement, people occupied large tracts of land without tenure; village settlements were few and far between; dwellings were often temporary; and livelihoods were generally sparse. A group of Baster farmers used the lands around the Sak River at what was to become the Rhenish mission station at Amandelboom. And in the early 1800s, small groups of Xhosa from the Eastern Cape had also crossed into the interior through this area, motivated by trading interests (mainly in ivory and guns), and/or seeking independence from Xhosa chiefs in the coastal inland (see Kallaway, 1982; Legassick, 2016). Elisabeth Anderson (1985) notes three groupings of Xhosa in the northern Cape area at this time: one on the southern banks of the Gariep/Orange river at Prieska from 1795; one at Pramberg established in 1809; and one at Schietfontein in the early 1800s.

In the Kareebergen, where the springs of Schietfontein and Harmsfontein were located, several thousand San most likely lived in nomadic small bands before the movement of pastoralists into the area.[3] Anderson (1985) notes the impact of pastoralists as follows:

> it was only by the beginning of the nineteenth century that pastoralist groups – Boer, Bastard, Khoi, and Xhosa – really began to penetrate there. This movement, though slow, led to the development of a complex frontier situation in which the intruding pastoralists co-operated to destroy San predators, and when that threat had been removed, the frontier remained an anarchical area, with little effort made by the central government to impose any form of control.
>
> (p. vi)

Diaries in the Bleek and Lloyd collection contain references to San at Schietfontein, and rock etchings in the region speak to its symbolic significance for the San as well as depicting animals that they might have encountered there. San stories have survived in this part of the Karoo, echoing in the stories told by older Afrikaans-speaking Coloured people living there to folklorist Manuel de Prada Sampras. No San languages have survived here, but descendants of the San are part of the local Coloured community in this place, as shown by the research surveyed by Parkington, Morris, and de Prada-Samper (2019).

The formal settlement of the area around the Kareebergen may be dated to 1830, when the Cape government granted 98,000 morgen for a Xhosa settlement in the Kareebergen just beyond the colonial border, including the springs at Schietfontein and Harmsfontein and the farm, Rhenosterpoort. This area later included Carnarvon. The northern Cape border had been extended towards the Kareebergen in 1824, and the apparent intention of the land grant was for this splinter Xhosa group to act as a buffer against the San, widely feared for their skill with arrows and their determined resistance to settler intrusion. Though just outside of the colonial border, this Xhosa settlement would nonetheless be accessible to government control.

The first leader of the Xhosa settlement at Schietfontein was Claas Hendrick, whose son, Jan Hendrick, changed his name to 'Jan Kaffer' (perhaps to signify his difference from and antipathy towards the San) and was sometimes referred to as *Kaptein* or Captain. He was succeeded by his son Daniel Hendricks. Although Jan Kaffer was initially not in favour of Basters being included in the Schietfontein settlement, over time Baster families intermingled with the Xhosa group. These families tended to be wealthier and have more stock. Given the arid conditions, people did not necessarily 'live' in the settlement, instead using it as a base as they moved their herds. Anderson notes that the settlement was seasonal from the start and that the Xhosa living there 'were nomadic and lived in small units over a 30 mile radius' – though by 1847, there were 'at least 620 Xhosa in the settlement' (Anderson, 1985, p. 45).

The patterns of interaction, settlement, and transhumance in spaces that had been marginal to colonial interests changed swiftly and dramatically in the 1840s with the introduction of merino wool farming into the areas around Beaufort West and Graaff-Reinet. Land that the colonial office had regarded as not worth the expense of measurement suddenly afforded commercial prospects to wealthier white farmers, whose interest was sparked in gaining large tracts of land for grazing merino sheep that were tolerant of the dry conditions. In 1848, the Cape Colony's frontier was extended northwards to the Orange/Gariep River, which meant that the Xhosa settlements at Prieska and Pramberg, as well as the Xhosa/Baster settlements at Amandelboom and Schietfontein, were brought under colonial rule.

Although the newly incorporated frontier area was officially deemed to be unoccupied, all of the suitable grazing lands were already in use by Xhosa, Basters, and poorer white pastoralists. Pressure on resources was intensified by frequent droughts in the area, as well as the general effects of frontier war in the Eastern Cape. It was in this context that a new politics of land claims played out. As will be seen, the interests of commercial farmers prevailed at the expense of marginalised groups already living nomadically in the area, and colonial regulations on land and movement of people were put in place – a colonial governmentality that fundamentally changed existing relationships of people to land and to each other.

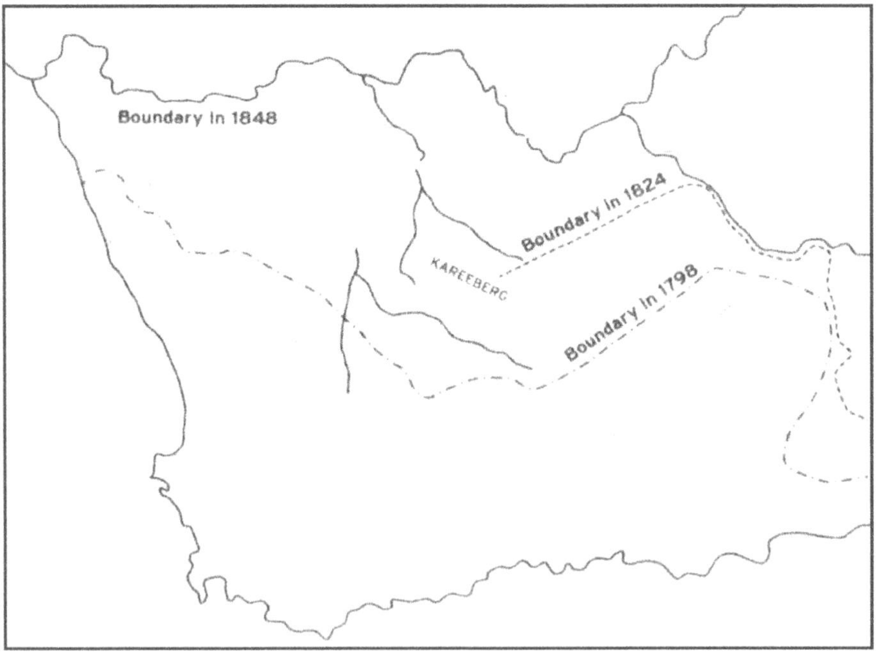

Map 3.3 Expansion of colonial boundaries.
Source: After Legassick (2016).

Enter the missionaries

Into this scenario stepped the Rhenish Mission Society (RMS) in 1847, with the intention of establishing a mission station to serve the Xhosa communities in the Kareebergen. Missionaries introduced another dynamic into this particular 'fragment of space', with significant and long-term effects. The RMS was a relatively small player in a large field of missionary endeavour at this time. Motivated in part by religious revivalism in Europe as well as Enlightenment ideals and humanitarianism, thousands of missionaries from Europe had set out across the globe in the 1800s, with South Africa being a popular destination. Richard Elphick (2012, p. 18) notes that there were no fewer than 385 mission stations in South Africa by 1884, associated with 15 substantial mission societies (the RMS being one of these). The RMS had been formed in 1828 as an association of three different German reform churches, operating from headquarters in Barmen, Germany.[4]

There is an extensive literature on missionary activities and their association with colonialism, the detail and debates of which lie beyond the reach of this study. Certainly, their active presence needs to be considered in analysing colonial spatial practices. Whatever their involvements beyond evangelisation, there is truth to Barker's (2005) observation:

> One cannot begin to make sense of the complex relationships between missionaries and other whites beyond colonial frontiers without bearing firmly in mind that the missions were, first and foremost, religious institutions. At the simplest level, the core functions of even the larger mission settlements centred upon classrooms where the essentials of the faith were taught and churches within which the faithful demonstrated their submission to God. ... The missions were only one face of a religious movement that was largely concerned with moral reform and spiritual awakening at home.
>
> (p. 100)

It is indisputable that, for missionaries, Christianity included western cultural practices, as Etherington, Harries, and Mbenga (2012) point out:

> almost all missionaries insisted that Christianity came as part of a total cultural package. Conversion must be accompanied by outward and visible signs of inner change: European styles of dress; a new kind of family life based on radical changes in ideas about masculinity and femininity; novel notions of health and hygiene; and adaptation to an economy based on wage labour, and an aspiration to accumulate capital.
>
> (p. 364)

It goes without saying that all of the communities in which missionaries operated were themselves involved in complex power dynamics of colonialism, in which local people were not simply passive recipients but also actors intent on shaping conditions to their own advantage or viewpoints. Missionaries aiming to westernise and Christianise local people were inevitably entangled in activities that disrupted local cultural practices and authority structures, and they were also involved in trade and economic exploitation. Yet – much as they were players in colonialism – it would be incorrect to read too much intentionality or even effectiveness into their activities, or too much passivity into the responses of local communities to their presence. Missionaries were not always successful in achieving their goals of evangelisation; there were often tensions between missionaries and colonial authorities; local inhabitants were not always obedient recipients of missionary endeavours; and local people were at times able to use missionary presence to further their own interests. This is certainly evident in the case of the RMS missionaries and the local people at Schietfontein.

In the ambiguous politics that had evolved around missionaries in South Africa, the establishment of a mission station at Schietfontein had the support of some of the local Xhosa leaders as well as local white farmers. Jan Kaffer and some of his group hoped that a literate advocate might serve their cause, while white farmers hoped that a mission station might curtail the freedom of movement of the Xhosa. (Ironically, both turned out to be the case.) The first missionary, Reverend Christof Alheit, located the mission settlement at the Harmsfontein spring, establishing a church of reeds and mud there in late 1847 and opening a school in early 1848. The Schietfontein spring, which was stronger, was designated for farming use. Not all of the Kareeberg Xhosa were supportive of hosting missionaries and some left the area to move northwards for alternative grazing spaces. Alheit's intention was to consolidate the different Xhosa settlements around the mission; however, he was soon taken up by a struggle against white farmers who were moving into the area, and he concentrated his efforts on ensuring that the Xhosa living around the mission retained the rights to the land they were occupying.

The battle for the land

Following the extension of the northern border of the Cape in 1848, the Colonial Office introduced new regulations for the tenure of Crown Lands, intending to formalise legal occupation of the interior lands and dispose of neglected areas in between formal holdings (see Anderson, 1985; Christopher, 1976; Legassick, 2016). Where Crown Land was already occupied as leasehold, it could be granted free of charge to the occupants, who would then be responsible for paying the costs for the land surveys and title deeds. Provision was made for Crown Land between private properties to be attached to one or the other at an agreed price. What followed was the rapid sale of land.

Missionary Alheit hoped that he would be able to use these new regulations to secure legal tenure of the land occupied by the mission station, since this had been granted for Xhosa settlement earlier on. However, this meant a struggle against existing governmental arrangements, since the British Colonial Office and authorities in Cape Town had little interest in the affairs of the northern frontier beyond curbing expenditure there, and instead were content to devolve land matters to local authorities whose bias was towards white farmers. Anderson (1985) provides a picture of the ugly local politics that ensued around land acquisition, with capitalist farmers (whom she terms 'the wool men') determined to acquire land by any means, including intimidation and duplicitousness (p 115). She argues that members of divisional councils and local authorities were complicit in malpractices, and also used opportunities to enrich themselves and their families. The result was that poorer white farmers, as well as Xhosa and Basters, were shut out of possibilities for acquiring the land in this part of the country once its commercial value had shot up.

As land was surveyed and allocated, the Pramberg Xhosa lost their land to white farmers and in 1855 their settlement was removed to Schietfontein. In a parallel contestation at the Rhenish mission station at Amandelboom on the Sak River, *trekboere* were able to displace Basters from the water sources where they had been farming productively for some time, and with the disintegration of the community, the Rhenish Mission Station later closed down there in 1874. Many of the Xhosa and Basters who were displaced by these changes in land regulations moved further into the interior towards the Gariep, seeking grazing areas. Others moved to Schietfontein.

With white farmers allocated the land around Pramberg and Amandelboom, their focus for land acquisition turned to Schietfontein itself. Seeking a suitable site to establish a town and build a church, these farmers viewed the spring at Schietfontein as an attractive possibility. Without consulting the existing inhabitants, they began to make plans to carve it up between them. By this time, in the late 1850s, the settlement at Schietfontein had grown in size to more than 1000 people with the incorporation of displaced Xhosa and Basters from the other settlements (Table 3.1). Pressure on the land was intense.

In 1857, when missionary Alheit learned of farmers' interest in claiming Schietfontein lands for themselves, he indignantly took up the fight against both the local divisional council and the Cape colonial government. Coming from Europe, he would have had some knowledge of land enclosures and their consequences, and his goal was to secure the Schietfontein land for the original Xhosa settlement as well as extra land around it for the additional Basters and Xhosa who had relocated there, in particular from Pramberg. He even attempted – unsuccessfully – to find alternative land for the mission settlement to the north at the Gariep River. In response to his approach, in 1858 the colonial government sent a surveyor, Auret, to investigate the various claims to the land in the area. His written report (hereafter referred to as Auret's Report), presented to the Cape Parliament in 1859 and located in its archives, gives insights into relationships of people, livestock, and land, and is therefore worth considering in some detail.

Table 3.1 Numbers living at Schietfontein mission station, 1859

Statistical return of the Missionary Institution at Schietfontein

	Number of families	Number of souls	Stock belonging to the respective families			Remarks
			Horses	Cattle	Sheep and Goats	
1. Kafir families at Schietfontein............	53	350	200	450	16,000	
2. Families removed from the Prambergen to Schietfontein in 1855	67	400	250	500	12,000	Many of these people have wagons and carts – the exact number Mr Alheit could not furnish.
3. Bastards and other families attached to the Church at Schietfontein..........	62	536	420	950	40,000	
Total....	182	1346	870	1900	68,000	

Source: Auret's Report.

Auret visited the lands around Schietfontein and Pramberg, investigating occupation patterns, numbers of animals, pasturage conditions, and water sources. In addition, he held a public meeting in the Schietfontein schoolroom to hear claims and objections from local people. His report states that 'a large number of farmers from all directions had arrived there' to meet him, and that the meeting was attended by 'about fifty farmers' and 'about one hundred and twenty Kafirs and Bastards belonging to the [Schietfontein] Institution' (1859, p. 13). Auret confirmed that there was no dispute among those present that land around Schietfontein, Harmsfontein, and Rhenosterpoort had previously been allocated to Jan Kaffer's Xhosa group. Place names mentioned in the Report as part of the original allocation were Meidefontein, Harmsfontein, Schietfontein, Boezaksfontein, Blaauwbosch, Minaskolk, Rhenosterpoort, and Kareekloof. However, white farmers present at the meeting made representations to Auret that Harmsfontein, plus a limited surrounding area, should be granted to the Xhosa from Schietfontein and Pramberg, and that Schietfontein should be made into a village

> for the benefit of the surrounding farmers, it being the best locality for the purpose in that part of the country, where, owing to the remoteness of the nearest village (Victoria) the establishment of a church for a white congregation ha[d] become very necessary.
>
> (1859, p. 14)

In compiling his recommendations, Auret attempted to ascertain the size of the resident population and quantity of livestock in the mission settlement and to make allowance for Pramberg Xhosa. Although he listed the details of Basters living at the mission, noting them to be ‘men of some substance’ in terms of the stock they possessed, he did not consider them to have the same land claims as the Xhosa. His report (1859) states that ‘these people, properly speaking, have no claims on the Government, similar to the Kafirs’ – even though he considered them ‘generally an orderly, well-behaved class of people’ and was ‘of opinion that some provision for permitting their continuance at the institution would be advisable for their own sakes, and, from their admixture with the Kafirs, have a beneficial effect upon the latter’ (p. 17).

Auret’s Report provides lists of Xhosa and Baster occupants of Schietfontein, and also lists of their stock. It is interesting to note that ‘mission’ or Christian names are recorded, signalling that Xhosa membership of the mission station implied the loss of Xhosa names. From names alone, the Xhosa and Baster members are indistinguishable to an outsider. It is interesting to note a smattering of women’s names among the men’s on the Xhosa list but not that of the Basters (Figures 3.1 and 3.2).

Auret recommended that the 200 erven he surveyed around the mission station at Harmsfontein be allocated as follows: 47 to Schietfontein Xhosa households, 65 to Pramberg Xhosa households, 55 to Baster households who belonged to the mission, one to the Rhenish mission, and 32 for sale by public auction with the proceeds to go to the mission. However, he recommended that only Xhosa erf holders and the mission itself were to be allocated ‘garden plots’, since this was the maximum that the Harmsfontein water could sustain. Every erf holder would have specified grazing rights on an inner commonage of 1200 morgen surrounding the village (which included Schietfontein), but only Xhosa erf holders would have grazing rights on an outer commonage of 86,000 morgen. This was a fraction of the land needed for the survival of pastoralists in this area, and stock quantities for the inner and outer commonage were to be strictly limited. In addition, erf holders were to erect a ‘substantial building’ to the value of £25 within three years (later changed to five years) and erven could be sold after five years. Until the establishment of a municipality in the area, Schietfontein was to be regulated by a Management Committee consisting of three Xhosa and two Baster members; each erf would be taxed; and quitrent would be paid to the government. This would contribute to the costs of a salaried superintendent with, in the words of the report, ‘the Government adding such an additional allowance as will secure the service of an efficient European, or other white person’ (1859, p. 18).

The colonial government accepted Auret’s Report, and in 1860 made two grants of land, one for an inner commonage to erf holders of Harmsfontein/Carnarvon, and the second for an outer commonage for grazing purposes, specifically to Xhosa erf holders from Schietfontein and Pramberg.

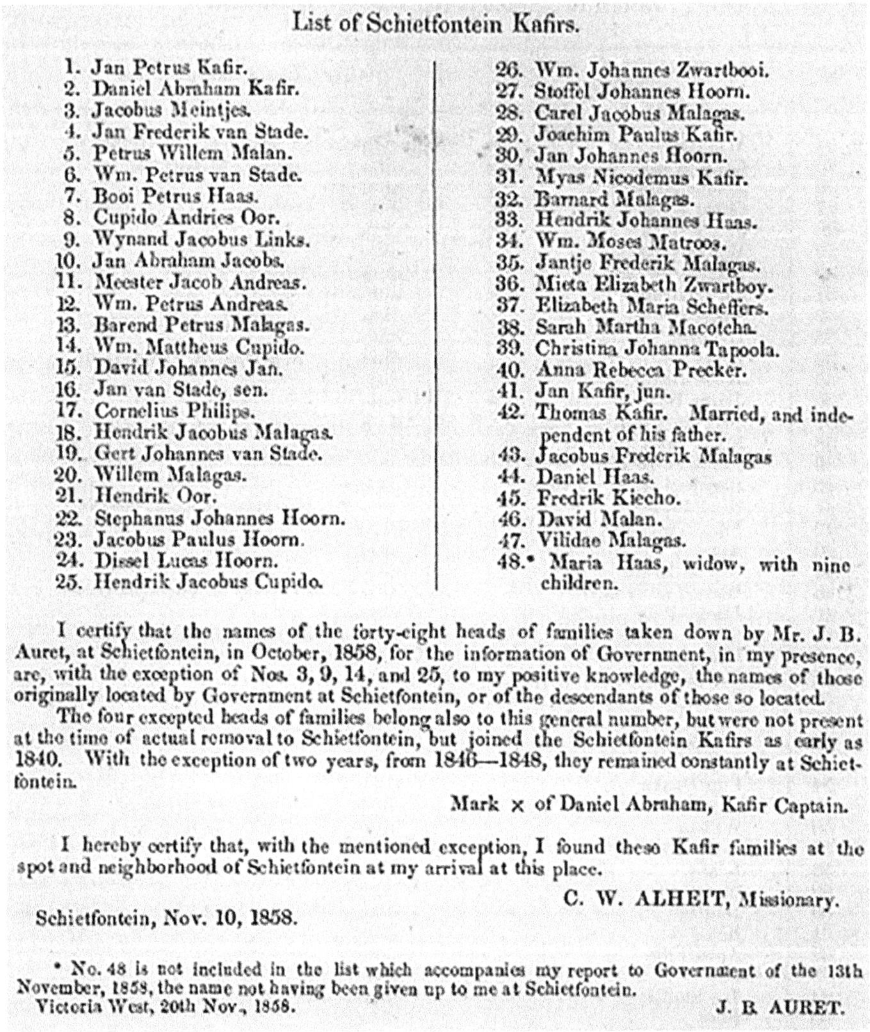

List of Schietfontein Kafirs.

1. Jan Petrus Kafir.
2. Daniel Abraham Kafir.
3. Jacobus Meintjes.
4. Jan Frederik van Stade.
5. Petrus Willem Malan.
6. Wm. Petrus van Stade.
7. Booi Petrus Haas.
8. Cupido Andries Oor.
9. Wynand Jacobus Links.
10. Jan Abraham Jacobs.
11. Meester Jacob Andreas.
12. Wm. Petrus Andreas.
13. Barend Petrus Malagas.
14. Wm. Mattheus Cupido.
15. David Johannes Jan.
16. Jan van Stade, sen.
17. Cornelius Philips.
18. Hendrik Jacobus Malagas.
19. Gert Johannes van Stade.
20. Willem Malagas.
21. Hendrik Oor.
22. Stephanus Johannes Hoorn.
23. Jacobus Paulus Hoorn.
24. Dissel Lucas Hoorn.
25. Hendrik Jacobus Cupido.
26. Wm. Johannes Zwartbooi.
27. Stoffel Johannes Hoorn.
28. Carel Jacobus Malagas.
29. Joachim Paulus Kafir.
30. Jan Johannes Hoorn.
31. Myas Nicodemus Kafir.
32. Barnard Malagas.
33. Hendrik Johannes Haas.
34. Wm. Moses Matroos.
35. Jantje Frederik Malagas.
36. Mieta Elizabeth Zwartboy.
37. Elizabeth Maria Scheffers.
38. Sarah Martha Macotcha.
39. Christina Johanna Tapoola.
40. Anna Rebecca Precker.
41. Jan Kafir, jun.
42. Thomas Kafir. Married, and independent of his father.
43. Jacobus Frederik Malagas
44. Daniel Haas.
45. Fredrik Kiecho.
46. David Malan.
47. Vilidae Malagas.
48.* Maria Haas, widow, with nine children.

I certify that the names of the forty-eight heads of families taken down by Mr. J. B. Auret, at Schietfontein, in October, 1858, for the information of Government, in my presence, are, with the exception of Nos. 3, 9, 14, and 25, to my positive knowledge, the names of those originally located by Government at Schietfontein, or of the descendants of those so located.

The four excepted heads of families belong also to this general number, but were not present at the time of actual removal to Schietfontein, but joined the Schietfontein Kafirs as early as 1840. With the exception of two years, from 1846—1848, they remained constantly at Schietfontein.

Mark × of Daniel Abraham, Kafir Captain.

I hereby certify that, with the mentioned exception, I found these Kafir families at the spot and neighborhood of Schietfontein at my arrival at this place.

C. W. ALHEIT, Missionary.

Schietfontein, Nov. 10, 1858.

* No. 48 is not included in the list which accompanies my report to Government of the 13th November, 1858, the name not having been given up to me at Schietfontein.

Victoria West, 20th Nov., 1858. J. B. AURET.

Figure 3.1 Copy of list of Xhosa at Schietfontein showing how they were recorded.
Source: Auret's Report.

There can be no doubt from even a cursory reading of Auret's report that divisions between population groups at this time were taken for granted and viewed as the legitimate basis for differential rights. It would be completely incorrect to overlook the racialised basis of social differentiation on the part of the colonial government, to say nothing of local community members, from the earliest times of the establishment of this frontier town. Auret's Report shows that the distinction between Basters and Xhosa was firmly drawn, with different status recorded. Auret (1859) recognised that Basters would not have sufficient land for pastoral pursuits, but envisaged that they

[E]

List of Schietfontein Bastards, &c.

No.	Name.	Date of joining the Institution.	Remarks.
1	Abraham Johannes Armstrong	1849	Late slav .
2	Fredrik Jacobus Viviers	1848	
3	Bartholomeus Koopman	1847	
4	Adonis Johannes Loth	1850	Late slave.
5	August Jacobus Bimidi	1851	Late slave.
6	Carolus Johannes Willems	1855	
7	Klaas Willems	1854	
8	Jan Abraham Paul	1853	
9	Jacobus Johannes Schulkwyk	1851	
10	Matthys Daniel Rooi	1850	
11	Marthinus Christian Klein	1850	
12	Fredrik Samuel Fortuin	1852	Late slave.
13	Barend Johannes Koopman	1853	
14	Arie Jephta	1851	Late slave.
15	Hendrik Pieters	1855	Late slave.
16	Hermanus Maasdorp	1855	A white person married to a colored woman.
17	Abel Fortuin....................	1853	Late slave.
18	Willem Johannes Cornelius	1851	
19	Klaas Jacobus Visagie...........	1850	
20	Hermanus Jacobus Duimpie	1850	Late slave.
21	Willem Johannes Van Wyk	1853	
22	Petrus Jacobus Duimpie..........	1853	
23	Cornelis Meintjes..................	1855	Late slave.
24	Jacob Mattheus Titus	1849	
25	Hans Van Wyk	1856	
26	Hendrik Jacobus Zwarts..........	1850	
27	Willem Maasdorp................	1854	
28	Hermanus David Titus	1848	
29	Jacobus Viviers	1856	
30	Gert Jacobus Johannes Pretorius ..	1853	A white person, arried to a colored woman.
31	Jan Bril	1854	
32	Hendrik Abraham Poolman	1847	Late slave.
33	Willem Carolus	1853	
34	Gert Julies	1850	
35	Gert Pretorius, jun.	1853	
36	Jan Van Wyk	1851	
37	Fortuin Mattheus Witbooy	1849	
38	Thomas Titus....................	1853	
39	Klaas Theron	1855	
40	Petrus attheus Danzer	1850	
41	Booi Dikkop	1854	
42	Arie Van Wyk	1855	
43	Jacobus Mattheus de Wee..........	1854	
44	Salomon Abdol..................	1850	
45	D vid Job......................	1849	
46	Jacobus Koopman	1847	
47	Ontong Johannes Snyman..........	1850	
48	Piet Van Wyk	1850	
49	Klaas Hendriks	1850	A Kafir, but not belonging to the old Schietfontein Kafirs.
50	Zwartbooy Umtea	1850	Ditto ditto.
51	Stuurman Zwartbooy............	1850	Ditto ditto.
52	Johannes Simon Slambo	1850	Ditto ditto.
53	April Matthews	1851	
54	Stoffel Sixteen	1852	
55	Willem Sixteen..................	1852	

NOTE 1.—Those who have more than one name have been baptized.

NOTE 2.—The above people, according to a rough estimate, possess jointly about 25,000 sheep, 300 horses, and 600 head of cattle.

Figure 3.2 Copy of list of Basters at Schietfontein.
Source: Auret's Report.

would 'nevertheless derive considerable benefit' from the right to own residential erven:

> by periodically residing in the village, with their families, as at present, for educational and religious purposes; whilst many of them who are able, by clubbing together, as I understand they intend doing, may purchase some of the other Crown Lands in the neighbourhood, which will shortly be sold; there being also room for others, for taking service with the surrounding farmers, as I am informed at present happens occasionally.
>
> (p. 18)

Similarly, the special provisions the report made for the residents of the new settlement indicate both the differentiation of groups and the potential antagonism between them:

> That the provision of the 11th section of the Act No. 24, of 1857, "for preventing Colonial Fingoes, and certain other subjects of Her Majesty, from being mistaken for Kafirs, and thereby harassed and aggrieved", be also extended to such of the Kafirs as have not been born in this colony, by the issue to them of certificates of citizenship; and that to such of the Schietfontein Kafirs who are colonial-born, and do not come under the provisions of the said Act, certificates be issued, stating that they are colonists, for their own protection in traveling, and for preventing other descriptions of Kafirs wandering about, and not provided with passes, from escaping apprehension on the plea of belonging to the Schietfontein community.
>
> (p. 18)

These arrangements were to prove decisive for the history of possession and dispossession in this place.

Possession and dispossession: land, municipality, and church

Land

Although the missionary settlement managed to retain the rights to the two springs at Harmsfontein and Schietfontein, plus Rhenosterpoort, for Xhosa and Baster occupation, against the wishes of the white farmers, the victory was pyrrhic and temporary. The land that was allocated was too little for pastoral livelihood, and it was wealthier commercial farmers rather than Basters who managed to secure the surrounding Crown Lands. A number of pastoralists who were not willing to reduce their stock left the area even before erven were allocated, and those who remained with reduced stock had few options beyond labouring on white-owned farms. As events unfolded, all of the allocated erven could not be taken up because people were too poor to finance the costs of the surveying and the required £25 building. In the crippling drought and depression of the 1860s, many of the original Schietfontein inhabitants who had taken up erven could not sustain the costs, resulting in further losses. The door was thereby opened for whites to legally purchase the land that had been exclusively allocated for Xhosa and Baster ownership, and to establish the town for themselves.

Thus, in effect, the arrangements that gave individual land ownership rights to Xhosa and Basters in this place simultaneously reduced the allocated landholdings to a level that could not sustain pastoralist livelihoods – and thereby, ironically, led to their dispossession and impoverishment.

A domino-effect of Xhosa and Baster dispossession followed the 1860 land grants. Land enclosure and private ownership – colonial governmentality – had the effect of shifting economic relationships and squeezing out those without means. Anderson (1985) notes that within ten years, Schietfontein had become a 'predominantly white village' (p. 126). While in 1857, there had been ten white children attending the mission school, by 1864 the number had grown to forty – this before a public school was established by the Cape Education Department. As the settlement of Schietfontein changed in nature from a mission station to a small town, the position of the Xhosa and Baster communities became more marginalised.

Marais (1957) suggests that until the 1860s, these communities 'did fairly well' in Schietfontein, and even better in neighbouring Amandelboom, but this did not last:

> In 1858 the 810 Xosas [sic] of Schietfontein possessed 28000 sheep and goats, and the 500 Bastards [sic] 40000, besides cattle and horses. New churches were completed in 1857 both at Amandelboom and Schietfontein (the latter valued at £1,300) by the sole efforts of the inhabitants, who were also contributing handsomely to the support of the mission, including the schools which served both White and Coloured. Yet by 1874 both Bastard and Xosa congregations had ceased to exist – at any rate as landholding communities.
>
> (p. 86)

Arguably, it was not the congregation of Schietfontein that ceased to exist, but its economic resources and social recognition. Incorporated as they were into private ownership in a monetary economy, members of the community were trapped into a vicious cycle of poverty and debt in terms of erf holdings, and by extension this trapped the RMS as well. In 1909, missionary Schmolke explained this to an RMS inspector as follows:

> According to law, these erven must not be rented, but must be cultivated by the property owner personally, so that if the owner suffered a loss of livestock through drought or other causes and could no longer

> support himself, the sale of the property is and was the only alternative. The buyer possesses no capital in order to pay for the inheritance, the station or community provides a loan, promises to make the interest payments on time, and then the story starts all over again. Only a few people have been able to hold on to their property and continue making the interest payments. The constantly changing ownership, financing of loans, and collection of interest payments has always been a source of difficulty and vexation for the station missionary.
> (Strassberger, 1969, p. 81. Translated by Dr Christopher Hedemann)

While the mission at Amandelboom was closed down, the station at Schietfontein continued, but it was no longer able to be self-sustaining as had been the intention of the RMS in establishing it. In 1882, the Carnarvon Commonage divided the outer commonage and grazing rights into *opstallen* (small farms) for the original erfholders, and over time, almost all of these passed into white ownership. In 1913, the inner commonage was divided among erfholders and in 1926, the Carnarvon Commonage Act declared the remaining land held in common by the Management Committee to be crown land, cancelling all existing communal rights. In spite of protests by people living on the commonage, they were rendered illegal squatters – on the very land that had been set aside for their forebears more than hundred years previously.

Anderson (1985) makes an interesting argument that Xhosa and Basters were not seriously concerned by the initial loss of erven at Schietfontein at the time of Auret's survey, since the application of Act 24 of 1857 (mentioned earlier) gave them equal citizenship rights to whites and they were hopeful of acquiring other Crown Land to the north. However, in the context of severe pressure to acquire Crown Land, she notes that racism began to feature in claims: 'northern frontier Boers who were unable to gain security of tenure and who were competing for land with the Xhosa and Basters now began to use colour as a sole reason for discrimination' (1985, p. 135). Boers reported that they found it 'demeaning' to have their ability to use land considered to be on a par with Xhosa, and race became an explicit means for whites to gain a competitive edge.

Kitshoff (1974), in his history of the *Nederduitse Gereformeerde Kerk (NGK)* in Carnarvon, presents clear evidence of the pattern of possession/dispossession that was underway. He notes that in 1876, 102 houses in the town were allocated: 35 owned by whites, 43 by Xhosa, and 24 by Basters (p. 28). Twenty years later, in 1894, a Rhenish mission inspector recorded that most of the Xhosa had sold their allotments, this during a period of terrible drought and impoverishment in the area. Commenting on the loss of Xhosa land tenure at Carnarvon, Kitshoff notes:

> As with their land, they also lost their language and identity. From 1934, they were no longer divided as a separate group in the Mission Church.

> They had actually blended in with the other coloureds and the language was Afrikaans, though a few people still tuned into the Bantu [sic] radio programme.
>
> (p. 152. Own translation)

Municipality

In its early days, responsibility for running the settlement was vested in the Management Committee representing erf holders. The handwritten minutes of the Management Committee (in English) record the early membership of Xhosa and Baster erfholders, with the superintendent of the reserve as chair and the resident missionary as secretary.[5] Records from the early meetings show three members signing with a cross, one of these being Daniel Hendricks, the son of the original Jan Kaffer, who is also recorded in the minutes as Daniel Kaptein/Captain (a term denoting leadership status). Other surnames that feature in the early minutes are Haas, Armstrong, and Witbooi (representing Xhosa erf holders), and Troon/Theron and Adriaanse (representing Baster erf holders) as well as the name of the chairman, Superintendent Dawson. Their descendants live in present-day Carnarvon.

However, as Xhosa and Basters sold their erven to whites for housing, administrative buildings, and business premises, the significance of the Management Committee reduced. Initially, its main tasks were to deal with matters relating to erven, water usage, grazing dues, and so on, but by 1868, it was heavily in debt. According to Anderson, it introduced money-making schemes of various sorts, 'the most lucrative being the sale of numerous spirit licences, hawkers' licences and a gunpowder licence for a private magazine' (1985, p. 132). In short, the Management Committee was dwarfed in significance as the white settlement grew. In 1874, the town was proclaimed a new magisterial district and in 1875 it was renamed Carnarvon after the British Colonial Secretary – an indication of growing British interest in the Cape interior as it opened up to commercial farming and then diamond mining after 1870. Having gained municipal status, a municipality was instituted in 1882, taking on the administrative functions of the town and eclipsing the Management Committee of the original erfholders.

Land loss implied loss of voice and communal rights for the original landholders. With the 1926 Carnarvon Commonage Act which ended communal landholding, the Management Committee was dissolved. At this stage, there were only two Xhosa families with full allotments in the town, and 16 with half allotments. In 1984, at the time of South Africa's Tricameral Parliament, 14 families identified as Xhosa, and they were classified Coloured (Kitshoff, 1974, p. 28).

The formal symbols of colonial governmentality were established along with the growth of the town: the municipality, the magistrate, the court, the prison, the bank. Traders and craftspeople with the necessary resources – often people from outside the area – were able to set up shops and small businesses to serve the town and nearby farms, often renting their initial sites from the Mission, which was struggling financially along with its members. By 1926 all of the land and institutions of the town were under white control – a zero sum arrangement of possession and dispossession under colonial conditions.

Church

In the early days of settlement, white settlers had been at least minimally prepared to share the mission facilities with local Xhosa and Baster inhabitants. Some attended the mission church (though they were catered to in a separate wing) and they sent their children to the mission school. However, as white settlement consolidated, white exclusivity in the institutions of the town became an entrenched practice. In 1874, the white community realised its hopes for a church. A congregation of the *Nederduitse Gereformeerde Kerk* (*NGK*) was formally recognised, and a new large church building signalled the status and serious intent of white Boer settlement. In 1880, a new and larger *NGK* church building was constructed around the corner from the existing Rhenish mission church. By this time, segregationist practices had already gained ground within the *NGK*, with the decision taken by the Synod in 1857 that racially separate services could be held. This was formalised by the Synod in 1881 with the establishment of a separate *Sendingkerk* (Mission Church) to cater for coloured congregations in the Cape (and later separate churches were established for missionary work among African and then Indian communities).[6]

In Carnarvon itself, once the *NGK* congregation was established, the assumption was that this would be a whites-only worshipping place. Kitshoff (1974) points to the reluctance of members of the *NGK* to reach out to the local Xhosa and Baster community, and the growing gulf between Afrikaners and others. He quotes from a letter that Sterrenberg, missionary at the Rhenish church from 1865 to 1892, wrote to his *NGK* counterpart pointing out that the white community showed no enthusiasm for missionary work among their own servants. Seemingly, the white community was unconcerned about the souls of the original inhabitants of the settlement they had taken over, and Sterrenberg noted that the gulf between the two groups had widened. A half-hearted endeavour by the *NGK* to provide a separate Sunday gathering for the Coloured community in the early 1890s was opposed by the Rhenish community and was soon abandoned. Much later, when the RMS withdrew from South Africa in 1943, negotiations were conducted for its Carnarvon church to join the *NGK* as a Coloured *Sendingkerk*, with its school as an *NGK* mission school – confirmation of segregation and inequality in the heart of the *NGK* since funding was not shared between the two congregations, which reflected the resource differential of their members.

Similar assumptions of segregation and 'ownership' applied to the schools in the town – as will be discussed in more detail in the following chapter.

Prosperity and poverty

In contrast to this picture of loss and dispossession of Xhosa and Basters, Kitshoff writes that the social and economic conditions for white farmers prospered, particularly with improved merino stock and good wool prices. Similarly, Archer (2000) notes that 'within three to four years of the diamond fields opening up in 1870, sheep prices in the Karoo doubled' (p. 683). Though the area continued to be plagued by droughts – some devastating – and the market for wool would rise and fall, capitalist farmers in the area were relatively well off, and the pattern was established of having a house in the town as well as on their farms.[7]

Archer (2000) outlines a number of significant changes to farming practices that extended economic opportunities for white farmers at the turn of the century. The introduction of fencing afforded free ranging of stock within camps, and ownership of land was consolidated into fewer, larger farms. Over time, the introduction of technologies for windmills and artificial watering also opened different spatial possibilities.[8] And the extension of the railway line (which included Carnarvon in 1905) brought access to markets and mobility to farmers, integrating the rural economy in new ways. One of the consequences of changes to land-use practices was that *bywoner* (tenant farmer) work was no longer as necessary as it once had been – though, as Lilja (2018) points out, shepherding remained important because of threats from jackals, despite farmers with jackal-proof fences replacing shepherds with camp walkers. (And when young adult workers moved from farms to mines, children and youth were used for shepherding tasks.) However, as capitalist farming expanded, the economic and social prospects of poorer people, white as well as black, continued to deteriorate. The 1890s were years of particular hardship for people with few resources to withstand the recurrent drought, disease, and pestilence. In this period, poverty stalked white communities as well as black, a point I return to in the next chapter in looking at the provision of schooling.

For those who had no land, regardless of race, rural life was harsh. As Colin Bundy (1984) notes in his writing on the Eastern Cape, various types of white poverty existed well before the officially recognised 'poor white problem' came into prominence during and after the 1890s to frame the issue differently. Bundy (1984) outlines the following conditions or types of poverty among whites:

> (a) the landless rural poor (*bywoners*, agricultural labourers, farm servants); (b) small-town low-skilled and low-paid wage earners (navvies, gangers, less skilled artisans, unskilled labourers, etc.); (c) a *lumpen proletarian* element (casual and infrequently employed labourers, a range of marginal and insecure quasi-rural employments, as well as drifters, beggars and criminals …).
>
> (p. 13)

Bundy also posits that relationships between these groups of rural poor were 'more fluid and interactive' in racial terms before the increasing racialisation of the 1890s. His archival research on the 1893 Labour Commission points to a good deal of racial mixing among the rural poor, and his research on the conditions of rural white poor in the Eastern Cape provides pictures of desperation in these records that warrant repeating: of 'unfortunate men … wandering about the colony almost naked and completely penniless'; 'skin and bones specimens of humanity'; 'emaciated forms from want are seen to move about like spectres from the tombs' (pp. 13–14).

Although Bundy's (1984) observations are based on the Eastern Cape rather than the Karoo, it is likely that landless poor people – white, Xhosa, Baster, remnant KhoeKhoe and San – would have struggled similarly for livelihoods on farms and in the town of Carnarvon. Low income or subsistence work would have been as servants, labourers, domestic workers, hawkers, and so on, and people would also get by through vagrancy, begging, and crime (p. 16). Bundy's research suggests that farmers would likely have treated this level of worker much the same, without racial favour. In terms of schooling, there can be little doubt that children with access to schools – including mission schools – would have enjoyed a level of privilege compared with their many counterparts who did not.

Summing up: settler colonialism established

The settlement of the area around Schietfontein and Harmsfontein in the early 1800s, and the establishment of the town of Carnarvon in the period after 1860, invite considerations about the nature of colonial power and governmentality as viewed from a marginal place in an intermediate time in South Africa's economic development.

First, though this may be an obvious point, the narrative of this particular place highlights the point that colonisation was not a monolithic process. The general colonial narrative, while holding true, is made up of many variations. Frontier dynamics in the northern Cape were not a replica of those in the eastern Cape and settlement of the interior showed different dynamics again. Land was settled differently according to geography and existing populations, and the forms taken by colonial violence were not identical. In this part of the colony, they included multiple smaller skirmishes and assaults as well as institutional incarceration and forced labour – not identical to the wars of the eastern frontier or the commandos of the north-west. Again, geography is likely to have been a determining feature. While proletarianisation of local inhabitants was a common outcome of colonialism, this was achieved in different forms. Such differences matter in that they inform local memories and leave specific dynamics and social relations in place over time. I suggest that part of the complexity of decolonisation in South Africa after apartheid results from local differences in governmental forms that complicate its dismantling. This is a point I return to in a later chapter.

A related point is that while San presence was recorded in this place before colonial occupation, the place was not intensively occupied, and their disappearance as a separate group from the area is indicative of their broader displacement through colonial violence. It is important to recognise, though, that individual San were incorporated into local settlements in various ways, though their distinctive presence disappears from official records. As Parkinton et al. (2019) argue, there is convincing archival evidence that San descendants are still living in the region. This is an issue of particular significance given that the Square Kilometre Array (SKA) uses a political grouping assembled elsewhere, the San Council, to acknowledge early San occupation, while completely overlooking their descendants in the local Coloured community of Carnarvon – as well as the historical ownership of the current Coloured community. I return to this issue in Chapter 9.

What is clear is that colonisation fundamentally shifted prior relationships to the land. The imposition of a capitalist logic of land enclosures and individual ownership was decisive in changing livelihoods. The spatial reconfiguration of the *trekvelden* lands also brought a shift in legal and administrative control of the space – a shift to colonial governmentality. Those without access to material resources and knowledge of the system were squeezed out in an increasingly market-based system of landholding and agriculture. As the area changed from being open *trekvelden* into private land ownership, people who had previously moved across the land found themselves without rights to the independent nomadic livelihoods that had sustained them. Possibilities of moving with stock ended forever, and lack of access to grazing land for stock inevitably meant proletarianisation for rural dwellers, white as well as black. Ironically in the case of Schietfontein/Carnarvon, it was the same legal instruments that simultaneously gave rights of ownership to Xhosa and Baster people and set financial conditions that excluded them from actual possession. Legislation linking erf ownership to financial resources meant that only those with money could acquire the sites originally designated for Xhosa/Baster occupation, and later it also enabled people to be evicted from the commons in the 1920s, resulting in their complete dispossession. This was not only a matter of access to financial resources to buy land; it was also a matter of access to the governmental 'rules of the game' – the use of law and information in capitalist ownership practices.

As possession of land hardened into patterns of exclusion, so the racialisation of these patterns became a taken-for-granted part of daily life in the social practices of the place. Racial classification, as a social construct related to power differentials, can be seen in early records of occupancy and access at both the mission settlement at Schietfontein and the town of Carnarvon, and this remained in different forms and intensities over time.

When looking at successive occupation of the land around the Kareebergen, what is striking is a sequence of formal displacements and exclusionary occupation in patterns of possession and dispossession: from San to Xhosa and Baster, from Xhosa and Baster to white farmers, and – as Chapter 9 will show – from white farmers to the SKA and its telescopes. The zero-sum patterns of possession and dispossession, established from the early days of formally recorded settlement, meant that forms of 'shared space' were segmented rather than held jointly or in common. Hegemonic assumptions of white entitlement and exclusivity, sealed by capitalist premises of ownership and the rule of law, inevitably meant that social relationships of class and race were fused to form deep structures of inequality, and no doubt this includes relations of gender as well. The 'sense of place' as a point of emotional connection was segmented from the start, without concern for building a 'common good' in connection to place, or nurturing of social bonds outside of employment relations between those living side by side. These are points I return to in relation to school restructuring in Chapter 8.

In terms of land ownership – a point that is crucial to rhythms and themes of possession and dispossession and political economy – it is important to consider the oral testimony gathered by *VG* church minister Sakkie Potgieter's (1999) *VG Kerk Carnarvon Gedenkblad,* a document celebrating the centenary of the Rhenish church. The testimonies, given by descendants of the families who had once owned property in and around the town, recall their parents' possession of farms as well as the names of the farms – names which match the farms recorded in Auret's survey of 1857.[9] They also recall times of dispossession, with their parents losing farms because they were unable to keep up with payments – presumably because of costs of various sorts, including fencing requirements, and the inability to raise capital to purchase allotments or erven.

In recent years, a land claim has been lodged for Schietfontein to be legally returned to its original owners, and it is likely that other evidence of ownership may yet emerge as the basis for land claims.[10]

Colonisation and identity

In looking at the patterns of relationships in Schietfontein (and later the town of Carnarvon), I suggest that it is important not to read the identity of Xhosa, Baster, and San in essentialist terms, or to assume tight links between 'identity', 'race', and 'tribe' – though, as I have shown, race was an active signifier in colonial governmentality, and racial prejudice was certainly evident. From the early days of its establishment, the mission settlement at Schietfontein catered to Basters as well as to Xhosa, and although the groups remained distinctive in part (assisted also by the classification imposed by Auret's Report), there was also intermingling within the community including marriages with San and with whites. Given that numbers of Xhosa had come and gone from the settlement at various stages, numbers are likely to have been small, and it seems unlikely that the traditional Xhosa cultural practices that remained would have been retained for long after the 1860 land allocations. Also, unlike the Pramberg Xhosa, those living at Schietfontein had been interacting with whites in the mission settlement for some years, which meant adopting different clothing and ritual practices, as well as adopting Christian names. Before that, they had interacted with white farmers and traders in this zone. Given that the names of those recorded as Xhosa are registered as Christian 'mission names', genealogy is difficult to trace through names, and it is impossible to trace through lifestyle and customs in this marginal community. Prominent families from the days of early Xhosa and Baster (and San) settlement include Boezak/Boesak, Sonn, Malgas, Cloete, Hoorn, and Daniels, and these names are still used in the town.

Very significantly in arguing against racial essentialism is the role allegedly played by Andries Boezak in the 1940s, when, as magistrate's clerk responsible for registering births, he apparently took the step of registering all births as Coloured.[11] While possibly motivated in part by reasons related to the church community, it was also the case that people classified as Coloured were accorded higher social status and potentially greater access to benefits than their 'Bantu' counterparts in the patterns of segregation that were being consolidated across the country even prior to the formal rules of apartheid.

Closing the colonial frontier: Anglo Boer/South African War

Before concluding this chapter, a brief comment on the South African/Anglo Boer War (1899–1902) is warranted, in that the war, as Wayne Dooling (2009) suggests, may be seen as 'the final phase in the closing of a violent colonial frontier' (p. 416.) It also brought a different moment of possession/dispossession for the town and the Rhenish mission school. During the war, the town fell under British martial law, and stock and supplies were requisitioned to support the British army. A British battalion was stationed at the town for a time. The mission church and its school were taken over as a British military hospital, with hospital tents set up on the church square and the *pastorie*/manse used as a residence for doctors and nurses. A blockhouse was built on the hill overlooking the town. Bill Nasson's (2011) comment about the blockhouses set up across the country as part of the British offensive is particularly apt: 'The mere presence of these little forts was an important stamp of imperial military authority' (p. 211).

A humiliating experience for the Rhenish community was that its German missionary, Reverend Stremme, was interred by the British on suspicion of being a Boer supporter (Strassberger, 1969).

Commenting on the involvement of indigenous people on the side of the British in the South African Anglo-Boer War in the northern Cape, Dooling (2009) notes that although white farmers might have succeeded in imposing brutal labour regimes on defeated indigenous people, they had not established their moral authority in what he terms a 'stunted agrarian economy'. Nasson (2011) provides an evocative account of Abraham Esau, a Coloured leader from Calvinia who worked with the British and was executed for this by the Boers, and there are other examples of involvement in the war to be found. Whether or not the local community in Carnarvon experienced the conflict in these terms is a matter for further investigation. What is known is that a number of whites from the area joined the Boer side and were later pardoned for treason after the war. Their photographs are to be found in the Carnarvon museum.

Wide-angle lens

After this first period of colonisation, we leave the municipality of Carnarvon firmly in the hands of Afrikaner residents, and with the original Management Committee of the reserve lands reduced in power and finally dissolved. We leave the church separated into a white-only *NGK* and a mission church catering to the earlier occupiers of the place – two separate buildings around the corner from each other, both in the same reformed church tradition. And, as the next chapter will show in detail, we leave the schools as separate institutions, funded differentially. Well before apartheid legislation locked down and amplified racial and spatial segregation, its major rhythms of practice had already begun to form in the colonial period in this place: a political economy which extended and consolidated ownership of productive assets in white hands and patterns of possession and dispossession which hardened racial divisions across multiple realms of political and social life.

Zooming out from a focus on this town in the Karoo to a broader picture of the geographical region in the 1800s, what is evident is that this was a definitive time of change more generally, with people from Europe penetrating into the lands and livelihoods of people living in southern Africa. There is a rich historiography that addresses the major strands of activity that shaped the status of colonial settlement at the Cape in this period, including the growing economy and its integration as part of the British empire into and the expanding world economy, the establishment of responsible government at the Cape in 1872, and the growth of administrative capacity and institutions.[12]

Duff sums up these major trends as follows:

> Before the 1850s, the Cape's value to the empire had lain entirely in its position on the sea route to India, but revenues from the export of wool and, after 1868, diamonds, transformed the Cape from a colonial backwater into a prosperous and significant settler colony located firmly within imperial networks of trade and administration. Industrialisation and economic growth fuelled the founding of towns throughout the colony's interior during the 1850s; the extension of hard roads and the railway across South Africa in the 1860s and 1870s; and the improvement of colonial infrastructure. The Cape gained its own Parliament in

> 1854, and was granted self-government in 1872. The second half of the nineteenth century witnessed the rise of a more 'modern' and industrialised state in the Cape.
>
> (2012, p. 263)

To this summary of trends may be added a number of others, set out as follows by Legassick and Ross (2012, p. 314): the prolonged and bitter conflict on the Cape eastern frontier, which ultimately resulted in the defeat of the Xhosa in 1853 and their incorporation into the Cape Colony as a proletariat (but not their cultural defeat); the 'Great Trek' in 1836 and the establishment of two fragile Boer republics in the interior and the Colony of Natal, amplifying hostility between Boer and British; and the complex relationships of class and race, which were increasingly hardening in hierarchical ways.

Viewed against these larger interpenetrating rhythms of activity, the story of the town returns to its marginal significance. It retreats in scale to be another small rural town in a sparsely populated, semi-arid place, with its own perceived-conceived-lived spatial practices.

At this point, it is appropriate to turn to the provision of schooling against the backdrop of the patterns of political economy, possession and dispossession considered thus far in this place.

Notes

1 Nell (2005) describes the name of this group (variously spelt) as follows:

> Bastaards were the Dutch-speaking, mixed-race descendants of the European-settler farmers, slaves, and Khoi who had been interrelating in the Northern Cape frontier region from the first decades of the eighteenth century. They held an ambiguous position in colonial society due to their mixed racial background, their adherence to Christianity, and their participation in the colony's militia. The Bastaard sons of settler farmers participated in settler institutions such as the commando system on more or less equal terms with colonists who were of wholly European descent. They also inherited, loaned, and purchased land in the region throughout the eighteenth century.
>
> (p. 123)

2 Christopher (1976) notes that the central Karoo and Namaqualand were largely avoided by pastoralists because of their aridity. By 1860, as little as a fifth of the land in the interior of the Cape Colony was held under land tenure (p. 78). Some isolated farms existed in advance of the pastoral frontier but much was unoccupied.

3 The most comprehensive primary-sourced account of this settlement is provided by Elisabeth Anderson's (1985) thesis entitled *A History of the Xhosa of the Northern Cape, 1795–1879*. Peter Kallaway (1982) addresses Xhosa presence in the area. Marais (1939/1957) includes sections on Schietfontein with particular emphasis on Basters, while Martin Legassick (2016) also mentions land dispossession of Schietfontein and surrounding areas in the Northern Cape. Xhosa presence in this area is also mentioned in Peires's (1981) classical study of the Xhosa polity. Activities of the three different major churches are addressed in Kitshoff (1974), Potgieter (1999), and Strassberger (1969), and Nel's (2015)

self-published text provides a local account of the history of the town. From their different perspectives, these sources point to the importance of the period leading up to and after 1860 in understanding how patterns of political economy, possession, and dispossession were established in this particular remote place.

4 The RMS Home Board in Barmen organized the placement, transfer, and recall of missionaries. The Barmen Home Board controlled the society's finances and also set out local structures for church councils and local conferences of missionaries in the colonial regions that it operated in, which included Borneo and Papua New Guinea. In South Africa, the RMS worked in Stellenbosch, Tulbach, and Worcester, and in more remote and isolated regions of the Cape northern frontier, such as Wuppertal and Concordia. Having already established a mission station among the Bastards at Amandelboom on the Zak River and with stations at Pela and in Namibia to the north, Schietfontein was the next choice. Before long, it had extended its activities across the border of the Cape in what was to become the German colony of South West Africa in 1880. Bade (1975) provides a fascinating account of the RMS's work.

5 Records of the Management Committee are held in the *NGK Argief*, University of Stellenbosch.

6 Whereas the church's formal position had been that it was open to all who were baptized and confirmed regardless of colour or status, the issue of how to include former slaves and black converts fed into debates about whether or not parishes and church services could be racially segregated. In 1857, the NGK Synod reached a compromise position: that parishes would not be segregated, but in cases where 'the weakness of some' (that is, people of colour) hindered worship, separate services could be held and separate facilities sanctioned. Whether envisaged at the time or not, the consequences of this 'Stockenstrom decision' in effect paved the way for segregation in the NGK more broadly. (See Giliomee, 2003; Ritner, 1967).

7 By the 1980s, landholding had consolidated into very large farms, and Nel (2015, p. 81) shows the number of farm owners dwindling from 348 in 1959 to 56 in 2015.

8 Archer (2000) mentions that these required considerable investment for individual farmers, requiring boreholes, boreholes, windmills, reservoirs, iron piping, and watering troughs. The introduction of windmills shifted farmers' dependence on surface water sources, opening different land use and spatial possibilities for those with money to invest in artificial watering, in order to raise productivity.

9 According to Potgieter (Interview, 2017) farms and owners were: Moordenaarskloof (Hoorn); Didaskloof (Andreas); Boezak (Lamoela); Witfontein (Pieterse); Minaskolk (Vass); Bruinheuwel (Janes); Rooileegte (Haas); Skurwekloof and Konka (no names); Swartbaadje (Links); Waaipunt (Petoors); Bloubos (Julius); Rooi Uitspanning (Jann); Xqinika (Malgas); Koerieskloof (Sukumbini); Berge (Haas).

10 Local resident, Sophia Vass, has provided the following list of land claimants: Martha Kiewiets, Sophia Vass, Jeanette Mathison, Antoinette Snyders, Andries Hoorn, Nicholas Beest, Louisa Hendricks, Gerty Julius, Hendriena Herandien, Peter Sawall, Wendy Erasmus, Hester de Wee, Marietha Pienaar. (Potgieter, personal communication, 15 January 2020.)

11 Interview, Louise Boezak, Carnarvon, 16 July 2017. The pages for this period are missing in the records of the Cape Archives.

12 Legassick and Ross (2012, pp. 289–290) refer to this as 'a major geographical expansion of European control within the subcontinent'. See also other chapters in Hamilton et al. (2012).

References

Official publications and archival sources

Cape Colony: Annexures to the Votes and Proceedings of the House of Assembly (Series Number G is by order of Government)

G7 - 1859 Report of an Inquiry into the Claims of Certain Natives Residing at the Missionary Institutions of Amandelboom and Schietfontein, in the Division of Beaufort, to the lands on which they are located. Presented to both Houses of Parliament by command of His Excellency the Governor, 1859 (Auret's Report).

Nederduitse Gereformeerde Kerk Argief, University of Stellenbosch

Sending Kerk (SK)

GS 1/1 Minutes of the Schietfontein Management Committee 1862–1877.

Secondary sources

Adhikari, M. (2011). *Anatomy of a South African genocide*. Athens: Ohio University Press.

Anderson, E. (1985). *A history of the Xhosa of the Northern Cape, 1795–1879*. Unpublished MA thesis, University of Cape Town.

Archer, S. (2000). Technology and ecology in the Karoo: A century of windmills, wire and changing farming practice. *Journal of Southern African Studies*, *26*(4), 675–696.

Bade, K.-J. (1975). Colonial missions and imperialism: The background to the fiasco of the Rhenish Mission in New Guinea. *Australian Journal of Politics and History*, *21*(2), 73–94.

Barker, A. (2005). Where missionary frontier ran ahead of empire. In N. Etherington (Ed.), *Missions and empire: The Oxford history of the British Empire companion series* (pp. 86–106). Oxford: Oxford University Press.

Brandt, F. (2014). Karoo parliament needs passion for radical transformation. *South African Civil Society Information Service*, 14 November 2014.

Bundy, C. (1984). Vagabond Hollanders and runaway Englishmen, white poverty in the Cape before poor whiteism, (pp. 11–23). Collected Seminar Papers. Institute of Commonwealth Studies, 33. ISSN 0076–0773.

Christopher, A. J. (1976). *Southern Africa*. Folkestone; Dawson; Hamden, CT.: Archon Books.

Deacon, J. (1986). 'My place is the Bitterpits': The home territory of Bleek and Lloyd's |xam San informants. *African Studies*, *45*(2), 135–155.

De Prada-Samper, J. M. (2012). The forgotten killing fields: 'San' genocide and Louis Anthing's mission to Bushmanland, 1862–1863. *Historia*, *57*(1), 172–187.

Dooling, W. (2009). Reconstructing the household: The Northern Cape Colony before and after the South African war. *Journal of African History*, *50*, 399–416.

Elphick, R. (2012). *The equality of believers: Protestant missionaries and the racial politics of South Africa*. Charlottesville: University of Virginia Press.

Etherington, N. K., Harries, P., & Mbenga, B. K. (2012). From colonial hegemonies to imperial conquest, 1840–1880. In C. Hamilton et al. (Eds.), *Cambridge history of South Africa: Volume 1 from early times to 1885* (pp. 319–391). Cambridge: Cambridge University Press.

Giliomee, H. (2003). The weakness of some: The Dutch Reformed Church and white supremacy. *Scriptura, 83*, 212–244.
Hamilton, C., Mbenga, B., & Ross, R. (Eds.). (2012). *The Cambridge history of South Africa, volume 1: From early times to 1885*. Cambridge: Cambridge University Press.
Kallaway, P. (1982). Danster and the Xhosa of the Gariep: Towards a political economy of the Cape Frontier 1790–1820. *African Studies, 41*(10), 143–160.
Kitshoff, M. C. (1974). *Kudde van Carnarvon: Nederduitse Gereformeerde Kerk, 1874–1974*. Carnarvon: N. G. Kerk.
Legassick, M. (2016). *Hidden histories of Gordonia: Land dispossesion and resistance in the Northern Cape, 1800–1990*. Johannesburg: Wits University Press.
Legassick, M. & Ross, R. (2012). From slave economy to settler capitalism: The Cape Colony and its extensions, 1800–1854. In C. Hamilton, B. Mbenga & R. Ross (Eds.), *The Cambridge history of South Africa: Volume 1 from early times to 1885* (pp. 253–318). Cambridge: Cambridge University Press.
Lilja, F. (2018). Inside the enclosed farm: Farmers, shepherds, and the introduction of new technology in Cape wool farming, 1865–1950. *International Review of Social History, 63*(1), 63–89.
Marais, J. S. (1939/1957). *The Cape coloured people, 1652–1937*. Johannesburg: Witwatersrand University Press.
Nasson, B. (2011). *The Boer War: The struggle for South Africa*. Gloucesterhire: The History Press.
Nel, de W. (2015). *Karoo kaleidoscope: A historic perspective*. Carnarvon: Self-published.
Nell, D. (2005). Treating people as men: Bastaard land ownership and occupancy in the Clanwilliam district of the Cape Colony in the nineteenth century. *South African Historical Journal, 53*(1), 123–145.
Parkington, J., Morris, D. & Rusch, N. (2008). *Karoo rock engravings: Marking places in the landscape*. Cape Town: Creda Communications.
Parkington, J., Morris, D. & de Prada-Samper, M. J. (2019). Elusive identities: Karoo |Xam descendants and the Square Kilometre Array. *Journal of Southern African Studies*. doi: 10.1080/03057070.2019.1647655.
Penn, N. (2005). *The forgotten frontier: Colonist and Khoisan on the Cape's northern frontier in the 18th century*. Athens: Ohio University Press.
Peires, J. B. (1981). *The house of Phalo*. Johannesburg: Ravan Press.
Potgieter, I. C. (1999). *VG Kerk Carnarvon Gedenkblad, 1947–1997*. Carnarvon: VGK Carnarvon.
Ritner, S. R. (1967). The Dutch Reformed Church and apartheid. *Journal of Contemporary History, 2*(4), 17–37.
Ross, R. (2012). Khoesan and immigrants: The emergence of colonial society in the Cape, 1500–1800. In C. Hamilton, B. Mbenga, & R. Ross (Eds.), *The Cambridge history of South Africa* (pp. 168–210). Cambridge: Cambridge University Press.
Rubidge, B., & Hancox, J. (2002). The Karoo supergroup: A geological and palaeontological superlative. *Rocks & Minerals, 77*(1), 54–59.
Skotnes, P. (1996). *Miscast: Negotiating the presence of the Bushmen*. Cape Town: University of Cape Town Press.
Strassberger, E. (1969). *The Rhenish missionary society in South Africa, 1830–1950*. Cape Town: Struik.

Interviews cited

Boezak, L. (2017). Interview, Carnarvon, 16 July 2017.
Potgieter, I. C. (2017). Interview, Carnarvon, 16 July 2017.
Potgieter, I.C. (2020). Email communication, 15 January 2020.

4 Schooling in place and time

The Cape Colony in the 1800s

Most accounts of the history of schooling in the Cape Colony give primacy to the activities of the Education Department. Its formal structure and the official format of its annual reports to the Cape Parliament invites a view of it as a recognisably modernist institution, managing the systematic expansion of schooling from Cape Town. However, if the vantage point is shifted from the governmental centre of the Colony to its outer frontier regions, this authoritative narrative gives way to a more uneven, less state-centric picture. Indeed, the picture shows the faltering extension of a frangible public system, heavily dependent on the support of churches to provide schooling. What also becomes apparent is that schooling, animated by different interests, afforded very different experiences in different parts of the Colony, and for different people.

The quotation below (taken from the official Blue-Book of 1841) provides a picture of schooling in the northern stretches of the Colony in the mid-1800s as seen from the official vantage point of the Department (with the terms 'inhabitants' and 'colonists' presumably designating white settlers):

> The great distance at which most of the inhabitants of this thinly populated Colony reside from any fixed Schools, obliges those farmers who can at all afford it to have private Teachers in their families; and in such cases the children of the neighbourhood generally resort to the residence of the Teacher for instruction, paying a small sum for his support. The education thus afforded is, however, necessarily limited, as well from the description of persons usually engaging themselves as tutors in the country as from the generality of parents being forced to avail themselves of the assistance of their children in their agricultural pursuits; nor has an instructor of superior qualification an opportunity of making himself useful beyond the family where he may be employed – the scattered state and indigent circumstances of the greater number of the colonists rendering it impracticable to form a School sufficiently large to remunerate him for his exertions. It is therefore (with the exception of the Missionary Institutions) in the several Towns and villages only, and at the expense of Government, that Schools can be kept up.
>
> (p. 251)

Shifting the vantage point from Cape Town to the Karoo itself, and more specifically to the different groups of people living around the springs of Schietfontein and Harmsfontein in the mid-1800s, the activities of the Education Department recede in significance. What becomes more apparent is the close relationship between schooling arrangements and settlement patterns, with local interests and activities giving shape to the forms of schooling that were established. From the perspective of the settlement in the Kareebergen, the first venture in schooling came from the Rhenish Missionary Society (RMS) in 1848, subsidised at a later stage by the Department of Education. And when the Department became more active in the northern frontier in the later 1800s, it relied very heavily on the support of the Nederduitse Gereformeerde Kerk (*NGK*) for its schools to operate.

What also becomes apparent when patterns of settlement are foregrounded as the vantage point rather than the activities of the Education Department is that schooling was not a prominent institution for shaping the lives of children – white, African, or Coloured – in poorer rural areas of the northern Cape during this period. While schooling as an institution was growing in importance in towns and wealthier white farming areas, the same was not true for places where populations were more dispersed. In such places, more common institutions for socialising children beyond the home were the labour market and the church.[1]

Before the introduction of merino sheep, pastoral livelihoods were sparse in the Karoo, particularly where people moved with their animals across the unenclosed *trekvelden* as grazing required. Later, as land was enclosed, those with farms enjoyed a certain measure of stability and prosperity, but those without access to land had neither. The discovery of minerals was to change the South African economy fundamentally, but in the times prior to this, most livelihoods were based on agriculture. Sometimes, farmers employed itinerant *meesters*, as described in the earlier quotation, and later on, small government-subsidised farm schools of questionable quality were set up. Where mission settlements existed, such as in Schietfontein, children living nearby could attend their schools. While wealthier farmers would have been able to send their children to boarding schools in the south-western Cape, this was not an option for most people on the northern frontier in this period. With the establishment of towns and their landmark *NGK* buildings, private and public schools were gradually opened for white communities, under the eye of both the Education Department of the Cape Colony and the *NGK*. But in the Karoo in the mid-1800s, most children received little schooling.

More commonly, children worked, not only on their parents' farms, but also as indentured labour on the farms of others, with boys frequently used as farm workers and girls as domestic servants. Under the Masters and Servants Act of 1841 and subsequent amendments, children could be indentured as labourers by their parents, with or without them, and 'destitute' or orphaned children could be indentured by local magistrates in loco parentis.

Lance van Sittert (2016a) argues that poorer parents indentured their children to strengthen their own bargaining position in the rural labour market, with children's labour being 'exchanged with settler pastoralists for subsistence, cash and/or livestock to subsidise, supplement and sustain adult household subsistence production' (p. 762).

From his analysis of statistics on child labour in the Great Karoo, van Sittert (2916b) estimates that there were 1,444 destitute children there between 1856 and 1909 (amounting to one-fifth of the colony's total). It would appear from his research that there were between 100 and 149 destitute children indentured in Carnarvon during this time (p. 42). While many of the destitute children in the Karoo were not classified by ethnicity, nearly half of those classified were 'Hottentot' and 'Coloured', and approximately a quarter were classified Bushmen, African, and Koranna. The remainder would have been white.

Bundy (1984) provides a striking description of what child indenture looked like, albeit drawing on research in the Eastern Cape rather than the Great Karoo. He states:

> Perhaps no more pointed illustration of the gulf between affluent and very poor whites is available than the workings of the legislation applying to destitute children. Under an Act of 1865, abandoned and destitute children were available for indenture, and were publicly advertised in local newspapers ... In 1876 a Mr Montague wrote to the Cradock magistrate in response to an announcement of three destitute white sisters:
>
> > I noticed yesterday for the first time, in the Cradock paper, that there were some destitute children shortly to be bound. I should much like to get one of the Girls, the one of 12 years old would be most useful, but if I can't get her the one of 8 would do ... Some three years ago I had a little English boy bound to me, & although at first I was inclined to regret of my bargain, I am beginning to find him very useful indeed.
>
> (p. 18)

In short, in looking at the development of schooling in the rural northern Cape Colony at this time, it is important to look past general histories of education to find a more complex and uneven mosaic of arrangements relating to childhood, work, and schooling.

The following sections outline the activities and worldviews – the telos or animating goal – of the three major actors in the provision of schooling in Schietfontein/Carnarvon at this time: the RMS, the Department of Education and the *NGK*. Each of these is considered first in its own terms, and the three are then woven together to show the trajectory of schooling in Carnarvon.[2] The analysis is underpinned by a Lefebvrean approach, which enables these different sets of practices to be analysed in terms of their own rhythms and flows as they interpenetrate and superimpose to constitute the social space of schooling. The arrangements put in place during this period were to have enduring effects, laying the basis for schooling for the next hundred years, and reverberating still in present-day Carnarvon, as I shall show.

Missionaries and schooling

As mentioned earlier, the first school in Schietfontein/Carnarvon was opened by the RMS in early 1848, conducted in the mud and reed structure that had been erected as the Church. The first teacher was the missionary himself, who was later joined by a teacher from Germany, and later still by locally trained assistants.[3] The first missionary, Reverend Christoph Alheit, drew up plans for a garden and encouraged people at the mission to adopt western dress and build houses for themselves. No doubt Alheit's wife supported his activities, particularly in organising activities for local women, and his wife and children would have considered themselves part of the broader RMS community. Elfrieda Strassberger's history of the RMS gives some attention to the Schietfontein station. She suggests that Dutch would have been a familiar language for RMS missionaries, but notes that mission work in isolated parts of the country was considerably more challenging than in more settled areas of the south-western Cape. The emphasis of the remote mission stations was on individual conversion and pastoral care, without haste to baptise. Central to the telos of the RMS was 'hard work', with mission stations expected to be self-supporting and even to contribute financially to the Barmen headquarters in Germany.

The educational work of RMS missionaries was second only to evangelisation in its importance, suggests Strassberger (1969), and as Rebecca Swartz (2019) notes, education in colonial contexts involved much more than formal schooling: 'Education involved transforming relationships to land and labour, and shifting religious and ideological positions of children, families and communities in varied social contexts' (p. 4). The priorities of RMS mission schools were Biblical history, learning Biblical texts by heart, reading, writing, and arithmetic – all within the broad ethos of 'combatting drunkenness and immorality' and a culture of work. The church trained local people as elders and deacons, and also built up a range of social activities to support its religious work. Involvement in the church choir, brass band, prayer meetings, girls' and boys' associations and the like became part of social life at Carnarvon and other mission settlements.

The report by land surveyor Auret, discussed in the previous chapter, leaves no doubt that formal instruction in the school as well as the church were enfolded in westernisation and the explicit adoption of a mission lifestyle at Schietfontein:

> I attended the church services for several successive Sundays, and estimated the average attendance at about 150 people, — consisting of Kafirs, Bastards, &c., in about equal proportions, and a few whites.

> The congregation generally were becomingly, and some even very respectably, dressed; and their apparent earnest demeanour on those occasions made a favourable impression on me. The discourses are delivered in Dutch, and very fluently interpreted by a native into the Kafir language.
>
> ... [M]uch good is silently working through the zeal and activity of the Rev. Mr Alheit amongst a class of people who are living in a remote and isolated part of the colony; and where, owing to the scattered nature of the population, and the original rudeness of the people, the labors for promoting their civilization must have been attended with more than ordinary difficulty....
>
> (Auret Report, 1859, p. 49)

Inevitably, since missionary activity is associated with colonialism, missionaries are often portrayed as 'agents of empire'. This is captured in sayings such as 'First they had the Bible and we had the land, now we have the Bible and they have the land'. Beyond these aphorisms, much has been written about missions and colonialism in different parts of the world, but comparatively little on mission education. Clayton Mackenzie (1993) provides a useful summary of the positions taken in the literature on this topic, including his own judgment:

> Missionary work in the last century alone spanned five continents and involved scores of different missionary societies and organisations. Two important 'mythologies' of missionary education have been identified as meriting closer consideration: (a) the mythology of missionary education as an arm of colonial conquest; and (b) the mythology of missionary education as an agent of social amelioration. These are deemed to be 'mythologies' because they represent expressed popular and/or academic archetypes of missionary educator functions and relationships. They do not purport to encapsulate the gamut of missionary mythologizing, but stand as important and valuable areas of debate. It will be argued that such archetypes, though partially valid, do not fully reflect the diversity and complexity of a wider and more balanced perspective ... [F]or the most part, missionary education cannot be viewed as a singular historical entity but as a series of disparate educational endeavours that encountered a variety of circumstances particular to their context and shared, to a limited degree and for differing reasons, a set of common experiences.
>
> (p. 48)

Norman Etherington (2005), a leading researcher in the field, points to the somewhat circumstantial links between missions and the British Empire, saying 'Just as the history of the British Empire can be written without much attention to missions, the history of missions can be written without much attention to the Empire' (p. 3). Making the general point more provocatively in the context of the RMS, Bade (1975) posits the challenge as follows:

> A critical analysis of the problem of colonial missions can neither be understood under the head of 'mission history' nor under that of 'colonial history'. One is rather forced to analyse the interdependence, consensus and collision of missionary and commercial, as well as political, interests.
>
> (p. 73)

Some missionary societies, including the RMS, were actively involved in trade and politics in certain parts of the world, depending on the affordances of context. While Bade's work shows the RMS's commercial and trading activities in mission stations in parts of New Guinea and their entanglement with German colonisation to the north of the Cape border, the poor circumstances of Schietfontien would have rendered commercial interests minimal, and indeed, the mission station there was not continuously self-supporting. In the sparse and stretched out spaces of the Karoo, not much economic benefit could be forthcoming for the RMS – and in the British colony, not much opportunity for German national expression. Nor could schooling in this remote place contribute much in terms of direct economic benefit.

In the Cape Colony of the 1800s, most missionaries were not British, and though they certainly promoted westernising worldviews, they did not necessarily support the activities of the British Empire. Most mission schools provided very rudimentary instruction to students in the process of attempts at evangelisation and westernisation. Although a handful of mission schools, such as Lovedale and Healdtown, were notable places of excellence, aiming to educate an elite in institutions that were comparable to schools in Britain, most mission schools provided very elementary teaching to those who had access to them, including white children. One of the contradictions of mission schooling, not only in South Africa but elsewhere also, is that many of the leaders in anti-colonial struggles were graduates of mission schools.

If mission stations are themselves viewed in Lefebvre's (1991) notion as 'fragments of space', it is possible to analyse multiple different flows of activity and meaning in their practices. These include: the European impulse to spread Christianity as expressed in the guiding vision of their particular mission society in Europe as well as its local structures; the different arrangements made with governments in colonies; the interactions with different local communities where mission stations were set up; and the subjective purposes, meanings, and practices of individual missionaries. And, as Achille Mbembe (2008) points out, the relationship between coloniser and colonised should not be seen as straightforward:

> the universalization of imperialism cannot be explained by the violence of coercion alone: it was a consequence too of the fact that many

> colonized people agreed, for more or less valid reasons, to become consciously complicit in a fable which they found attractive in a number of respects. The identity of the colonized and of the colonizer was shaped by the intersection between ellipsis, disengagement and renewal.
>
> (p. 3)

Strassberger (1969) periodises the RMS's work in Schietfontein/Carnarvon into three: 1845–1860; 1860–1902; 1903–1936. She describes these as follows:

> The first period was a time of prosperity and rapid growth; the second and third periods brought much suffering and economic misery, but the church work continued all the time in staunch faith despite adversity. During the third period the congregations remained mission congregations, until they were transferred to the D.R.C. [*NGK*].
>
> (p. 85)

Though economically struggling, the mission church in Carnarvon continued to draw numbers, as did the school – an important point to note in understanding the continuing presence of the church and school for the Coloured community through the apartheid years. However, important as the RMS's contribution was to the initial template of schooling in this place, it was the colonial government that increasingly provided the framework for schooling as white settlement increased in density. The next section provides a brief outline of this framework and the telos and activities that it entailed.

Department of education of the Cape Colony

Officially, schooling in the northern Cape fell under the aegis of the colonial government, and after 1839, its Department of Public Education and Superintendent General of Education (SGE) in Cape Town.[4] Having taken over the Cape from the Netherlands in 1806 and formally established its status as a British colony in 1814, the British colonial government proceeded with the anglicisation of administrative structures and government institutions, including schools. After a faltering start with free public schooling in the early years of the Colony, a new system of graded schooling with grants-in-aid funding was put in place after 1841. In 1865 an Education Act was passed, following the extensive Watermeyer Commission of Enquiry (1861–1863). The Act formalised the system of graded schooling and adjusted the grants-in-aid scheme to provide differential funding for non-denominational public schools based on the community capacity to support schools (basically a £ for £ system). In larger centres, provision was made for First Class schools with primary and secondary classes and a maximum grant of £200; in smaller centres there would be Second Class schools with primary classes only and a maximum grant of £75; and in rural districts there would be Third Class schools to provide minimum instruction with a maximum grant of £30. English was to be the medium of instruction in First and Second Class schools (this was changed in 1882 to allow for Dutch to be taught); religious instruction was to be offered but attendance was not to be compulsory; and schools were to be inspected. Separate funding provisions were set up for mission schools and 'aborigine schools', and private schools were also recognised.

Over the years, as white settlement extended into the interior, the Department took a number of measures to extend the reach of the schooling system in rural and remote areas. After 1872, additional inspectors were appointed to cover the expanding number of schools across the large geographical spread of the Colony. In 1873, a system of grants-in-aid was set up for district boarding schools to cater to white children living remotely.[5] In 1882 a per capita allowance was made for payment of private teachers on remote farms provided there were at least five students and a year-long programme; in 1887, a system of itinerant Circuit Teachers was introduced; and in 1893 this was replaced by the provision of free schools for poor white children.[6]

However, schooling was not compulsory, and the system of state aid depended upon community and/or parent initiative for schools to be established.[7] This in turn depended on whether or not the form of schooling provided matched the values and livelihoods of its intended subjects, as well as the degree of government presence in the area – which differed across the varying conditions of the Colony. The result, during this period, was low overall enrolment in schooling, intermittent attendance, and limited learning by the majority of those who attended schools beyond the larger settlements (Figure 4.1).

In the mid-1800s, Blue-Books and Department of Education Reports show an expansion in the number of schools, but also their uneven dispersal through the wide spaces of the Colony.[8] Schooling reached no more than a third of the white population at this time, and the standard was generally low. Again, in the concertina image, the system at its stretched-out parts bore little resemblance to its pushed-in parts in terms of quality and concentration of resources. From the vantage point of Cape Town and surrounding towns, the picture is one of the steady establishment of public and private schools and increasing enrolment of students. By 1860, there were more than 50 public schools in the Cape, with a number of First Class and private schools concentrated in the south-west. Sarah Duff (2015) provides an overview of the establishment of these preeminent schools: for example, the South African College in Cape Town (1829), Grey Institute in Port Elizabeth (1856), the Gymnasia in Paarl (1857) and Stellenbosch (1866), and Graaff-Reinet College (1866) (p. 93). These schools had extensive grounds and impressive buildings and offered comparatively high standards of schooling for wealthier families living in these more populated places, as well as boarding facilities for the children of wealthy farmers living further away.

However, opportunities such as these were not available to people living in more remote places or on farms, or to poorer people – as acknowledged

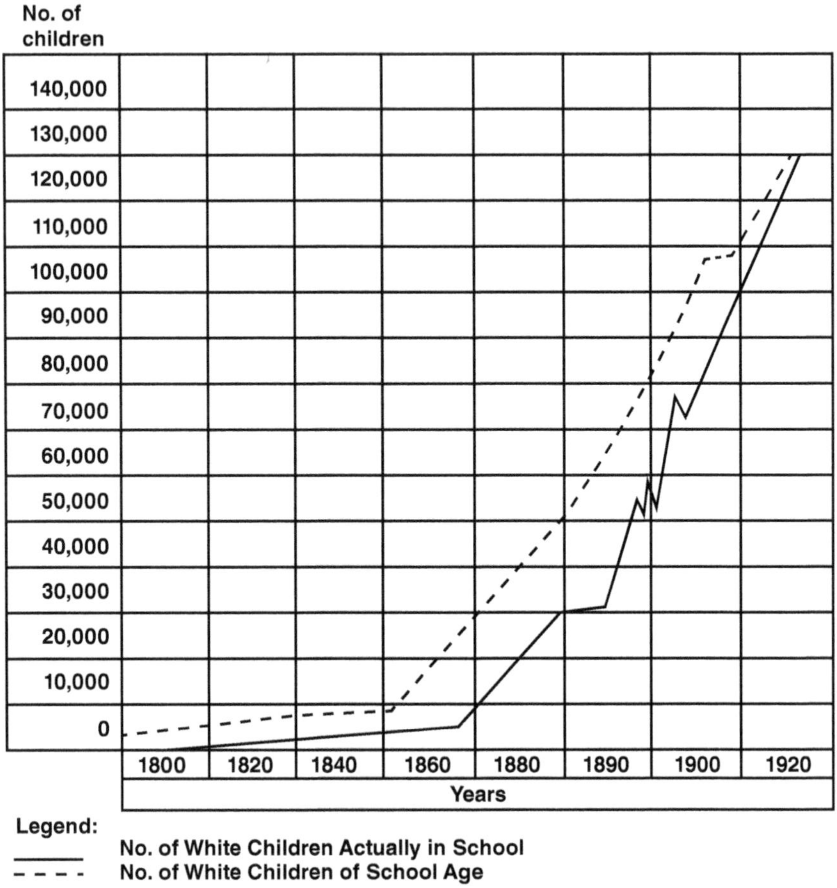

Figure 4.1 Growth in school attendance in the Cape Colony, 1800–1920.
Source: Malherbe (1925).

by the SGE's Annual Reports from this period, and illustrated by the quotation given earlier from the Blue-Book of 1841. SGE reports and those of several commissions suggest that this picture did not change much over the rest of the 1800s.[9]

Until the 1890s, government grants did not include provision or maintenance of school buildings, and inspectors' reports show that the buildings used for schooling in poorer areas and farms were often in bad condition. Inspectors' reports mention poorly lit, unventilated rooms with mud floors; furniture consisting of packing cases and parts of discarded farm equipment; and the school room being used to double up as accommodation for the teacher. More alarmingly, the reports show intermittent attendance and they lament the low levels of students' performance in reading, writing, and arithmetic.

SGE Dale's Report of 1887 was particularly concerned that the 'sons and daughters of the soil' (in other words, the children of white farmers) should develop the 'habits of continuing industry and perseverance' and 'love of work' that were necessary for success in a modernising economy. He saw the Cape as 'entering upon phases of life stirring, pushing and speculative, and upon new industries requiring energy, intelligence and well-regulated physical powers', and feared that farm children in particular would be left behind if they did not take up opportunities for education. However, the government's system of 'undenominational public schools' financed by grants-in-aid provided the most benefit to those who could afford better schools, and often not much benefit to poorer people, many of whom were left out completely. The fact that public schools were English medium for most of the 1800s and had British-oriented curricula would also have contributed to the sense that they were of marginal relevance to the lives of those outside of larger towns, the majority of whom spoke Dutch-Afrikaans.

As mentioned earlier, the Department relied almost entirely on mission schools to provide for black people and gave supplementary grants to them from 1841. Mission schools were not racially exclusive and included numbers of poorer white children, particularly in towns. While the Department included these schools under its regulatory framework, the SGE Reports were scathing about the quality of instruction they offered. In his Annual Report of 1861, Dale provided an unfavourable picture mission schools, and he quoted this in subsequent Annual Reports as well to express his concerns:

> It is my duty to record that in the numerous and well-attended mission schools of the colony, not more than five per cent. of the children reach the moderate standard of ability, to read fluently, to write fairly, and to work with facility and correctness examples in the usual simple rules of arithmetic. Three fifths of the children in these schools do and learn literally nothing, beyond the repetition of a string of biblical facts and yet there is an apparent satisfaction in the minds of the managers and supporters of these schools. In several schools the condition of aid which requires that the English language shall be a subject of instruction is evaded, and in many others, it is difficult to find even two or three children who can read either Dutch or English with ease and intelligence. … [A]t present the labours of the teachers are almost exclusively confined to the highest class, the other classes being committed to the charge of monitors, who, in most cases, are as ignorant as those under their care, and who in many cases, do the children more harm by their example than good by their instruction.
>
> (pp. 4–5)

Dale was particularly uneasy about the presence of white children in mission schools, complaining that this was inappropriate since their parents were not part of missionary congregations. He favoured, instead, the establishment of a Fourth Class Public School, since: 'It is very undesirable that the white children, especially girls, should be brought daily into close relations with the ordinary type of the coloured street boy' (1890, p. 7). Estimates are that up to a third of white children attended mission schools at the turn of the century.

That the provision of schooling was class-based and racially inflected was highlighted in the Preliminary Report to Parliament by the newly appointed Inspector-General of Schools and Colleges, Donald Ross, in 1883.[10] Ross, with his experience of the Scottish system, was highly critical of what he saw:

> It must be noted that the First Class Schools are to a large extent the schools of the aristocracy or ruling classes ... In the same way the 2nd Class Schools are more or less the property and the preserve of the superior ranks. ... I have ascertained that it is not uncommon to admit into all classes of schools the children of the poorer citizens at reduced fees, or even free, but obviously the arrangement is defective, and generally the poorer white population in towns send their children to inferior Mission schools where most likely they can learn to benefit them, or to private adventure schools, or they neglect them altogether. Those few people of colour who could afford did make use of the opportunities that came their way.
>
> (p. 12)

The Ross Report (1883) highlighted a number of weaknesses in the administration of the Department as well as the quality of schooling. Ross was particularly concerned about the disparity between the 'ideal side, as it appears in Blue-books' of the Department's work and 'actually, as it is to be found in its working-dress over the country' (p. 16). In his view, the Department was understaffed; it concentrated its energy at the top end of the system; and the level of instruction in schools was generally 'very low indeed'. He was particularly critical of what he regarded as the Department's 'unreliable or inflated' statistics and estimated that 'only one-sixth of the children of school age in the colony attended school with beneficial regularity, whilst five-sixths were outside the Government system' (p. 9). He remarked on the high number of very good private schools in the south-west, particularly those run by religious establishments – as well as the high number of 'adventure schools' at the other end of the spectrum. While commenting on the 'splendid buildings' of the First Class schools and a number of private schools, and praising girls' schools as generally better than boys', Ross found the overall system unsatisfactory: 'Briefly put, not much more than about one-fourth of the white population are in receipt of such an education as develops character or enables them to compete in the battle of life with those trained in Europe' (p. 15).

The Ross Report also highlighted the fact that most schooling did not extend beyond four years, and that students were concentrated in the lower classes, as the figures contained in his Report show (Table 4.1).

Table 4.1 Number of children at school and their grade levels

Number of children at school						
	Below	*I.*	*II.*	*III.*	*IV.*	*Above.*
European	5,362	2,784	2,721	2,550	1,118	181
Native	24,278	4,858	2,204	648	446	9
Total	29,640	7,642	4,925	3,198	1,564	190

Source: Ross Report (1883, p. 16).

What the Ross Report highlighted is that the Department was actually a slender structure, incapable of covering the broad geographic spread of the Cape. Until 1872, when additional inspectors were appointed, the SGE alone was responsible for the inspection of all schools as well as the administration of the system.[11] Although SGE Dale was understandably indignant about Ross's criticisms, it was nonetheless true that the Department relied on school managers to provide statistics of enrolment and attendance, and these were often inaccurate or simply not submitted.

A major factor was that the system of public schools supported by grants-in-aid was essentially a voluntary system, depending on the interest and capacity of local communities to provide schooling. Dale was a strong supporter of voluntarism in education, viewing the state's main role as being to remove hindrances and to support parental initiative, rather than to establish uniformity and control in a state system. As a result, schooling was very unevenly provided in the colony. School attendance was better in the larger centres, but in most of the colony, attendance was irregular.

This point was noted in the 1891 Commission of Inquiry report (Barry Report), which found that 'little more than one-third of the European children of school age in the Colony were in reputed attendance at school' (1891, p. 76).[12] Reasons given for this in the Report included the apathy and indifference of parents, the voluntary nature of the system, the lack of facilities, and the general levels of poverty in small towns and rural areas where children were required to work where possible and parents could not afford schooling. Indeed, the *Special Report* of 1901 notes in particular the 'purely voluntary nature of the system', its lack of planning, and the fact that 'No provision whatever was made for bringing new schools into existence at places where such were needed, or for securing the attendance of children within reasonable distance of an existing school' (1901, p. 74). In short, to use Ross's terms, the 'working dress' of the Department did not match its 'ideal side' as presented in official documentation.

In concluding this section on the Cape Education Department, a further comment on its governmental form is warranted. As mentioned earlier, the Cape Colony was the first of the British dominions to establish an education department (in 1839) and it is easy to assume an institutional coherence from its documentation. The presence of prominent and visionary Superintendents General of Education (SGEs), such as Rose-Innes (1839–1859), Dale (1859–1892), and Muir (1892–1915); the well-presented Annual Reports of the SGE to the Cape Parliament; and the official format of its enquiries lend themselves to a coherent governmental reading. The modernist discourse of the department suggests a narrative of systematic progress in the expansion of a coherent education system, unfolding over time under the control of its centre in Cape Town. However, Ross's warnings about the self-presentation of the Department and the accuracy of its statistics are worth heeding.[13] And certainly, from the vantage point of the northern Cape and Carnarvon in the mid-1800s, the authoritative governmental narrative gives way to a more uneven, less state-centric picture. What becomes apparent are the limits to governmental power and the importance of different churches in providing schooling. What also becomes apparent is the differentiated and uneven nature of schooling provision.

Schooling and the *NGK*

Given the towering presence of the *NGK* in South Africa before and during the apartheid years, it could be easy to assume that it had the same importance in the early years of the Colony as well. Yet its history should not be read backwards or as a simple unchanging trajectory. The influence of the church certainly extended back to the early years of Dutch occupation at the Cape, and under British colonialism the *NGK* was one of three subsidised churches. However, as Sara Duff (2015) points out, until the Great Revival of 1860 it was largely 'a socially and theologically moribund organisation'. It was in the process of the Great Revival that it became, in Duff's words, 'the very centre of, particularly, white Dutch-Afrikaans society in the colony's towns and villages throughout the colony' (p. 23).

Hermann Giliomee's (2011) comprehensive study, *The Afrikaners*, provides an intriguing depiction of Cape Dutch society and church in the 1800s.[14] Giliomee shows the emergence of a clear class divide between the wealthier colonial elite living in Cape Town and surrounding farming areas of the south-western Cape, and the Dutch pastoralists – generally poorer and less educated – who had moved into the interior. He points out that the Cape Dutch elite, whom he calls 'the Queen's Afrikaners', were increasingly identified with the Empire, participated in a civic life where English was the official language, and supported the British imperial narratives of progress and modernisation. Similarly, Robert Ross (1993) points out that the countryside of the Cape was dominated by a gentry who controlled land and labour, first in the wealthy wine and wheat farming regions of the south-western Cape, and subsequently in the wool farming areas of the interior. He speaks of a model of 'large, owner-occupied farms worked by a harshly exploited black labour force' (p. 48).

Susan Newton-King's (1999) *Masters and Servants on the Cape Eastern Frontier, 1760–1803* throws light on the living conditions of poorer cattle farmers or *veeboeren*. It is clear that rural Dutch/Afrikaners who were not landowners contributed to an underclass of tenant farmers (*bywoners*) who were without means and poorly educated. Duff (2015) notes that:

> the apparently ever-enlarging group of *bywoners* (tenant farmers), many of whom led a semi-nomadic existence, moving seasonally or as pasture deteriorated and water supplies dried up, was a cause of anxiety for the Dutch-Afrikaner middle class, who worried that their existence was a sign of Dutch-Afrikaans cultural degeneration.
>
> (p. 8)

Similarly, Bundy (1984) comments on the sense of social distance that well-to-do white farmers showed towards poorer whites in the eastern Cape: 'Certainly, in the 1870s and 1880s, a large number of spokesmen for the "haves" among the white population displayed a robust disdain and even hostility towards the "have-nots" in their midst' (p. 17).

Moreover, it is interesting to note with Giliomee (2011) that it was the British and European settlers (including Jewish immigrants) who came to the Cape during and after the 1820s who took up most of the opportunities that opened in towns and rural areas as the colonial economy grew and changed under British rule. Though they were often poor on arrival, they were more skilled than their local counterparts. Giliomee notes that 'it was they, rather than the Afrikaners, who first saw the potential of large-scale land speculation', they who introduced wool farming during the 1830s, and they who came to dominate trade and industry, particularly as the economy opened up after the mineral revolution (p. 194). It was only during and after the 1870s that Afrikaner organisations gained traction in political life.

Although most towns in the interior were established as *NGK kerkplekke* (church places) with funding from rich Dutch/Afrikaners, Giliomee notes that the church generally tended to be loyal to the government of the day and traditional in its ways. In the 1850s, however, the church itself began to change, due in part to the arrival of Scots ministers as well as younger and more liberal clergy who were influenced by their experiences in seminaries in the Netherlands. As well as a more liberal strand, different conservative strands also emerged within the Cape church. Among the prominent conservatives was Reverend van der Lingen, who established the Paarl Gymnasium as a Dutch-medium private school in 1857 – a forerunner to Christian-National Education, whose influence extended to the Carnarvon *NGK* and its school, as we shall see. In contrast, more moderate younger conservatives tended to view the colonial government in a favourable light. Among these was the highly influential Andrew Murray Junior, Moderator of the *NGK* from 1862 to 1897, who avidly supported the Cape Education Department in its drive to establish schools.

It was in these circumstances in the 1860s that the church experienced an outburst of evangelical enthusiasm, being struck by waves of religious fervour, piety, and emotional personal conversion. This Great Revival was linked in part to the rise of evangelism in Europe as well as different movements within the church itself. Interestingly, it was a religious movement that was not overtly political. Alongside the evangelical experiences of individuals, a host of activities flourished: special church services, Sunday schools, prayer meetings in small and large gatherings, religious clubs and societies, and a flurry of publications and periodicals. Duff (2015) shows how concerns about children and childhood moved into centrality in the church's discourse of this time. She suggests that Andrew Murray Jnr's statement '*Wat zal duch dit kindekin wezen*?', translated at the time as 'Raising Your Children for Christ', would be better translated as 'What shall this child become?' (foreword) – a statement which captures a pastoral preoccupation with (white) childhood within the revivalist movement and the *NGK* more broadly, particularly in its wealthier congregations of the south-western Cape. Unsurprisingly, this focus on childhood dovetailed with general concerns about education, with Moderator Murray and SGE Dale in close accord about the need for more schooling, as mentioned earlier. In 1873, the *NGK* appointed a special inspector to liaise with the Education Department.

Indeed, from the 1870s onwards, the *NGK* became the major player in the provision of schooling in rural areas. In Malherbe's (1925) words, the *NGK* was 'the very soul of local educational effort', and in his view the schooling system in the Cape would have collapsed without *NGK* support (p. 109). He notes that the *NGK* supported education throughout the country 'by sympathetic supervision, by establishing institutions of higher education (especially for girls), by actively supporting teacher training and industrial schools, and by bringing mother-tongue instruction to its rights in schools' (p. 117). The *NGK* became increasingly concerned about the illiteracy and generally deprived circumstances of poor white children, particularly in rural areas but also in towns. The role of the church in pastoral care was expanded, as well its role in schooling.[15] In the 1890s and subsequently, both church and state acted to support the education and care of growing numbers of poor white children, including the establishment of 'poor schools' in 1896.[16]

What, then, of language of instruction in these schools? While schools around Cape Town were generally accepting of English medium and the teaching of Dutch, the issue of language medium came into greater focus as schooling spread into the interior. By the mid-1870s, forms of simplified Dutch were increasingly used across the country, including what came to develop as Afrikaans, a partly creolised form of simplified Dutch used between masters and servants, white and black.[17] It was clear that outside of the elite of the south-western Cape, few people could read and write in Dutch, particularly since very few schools taught Dutch. Giliomee (2011) makes the following point about the development of Afrikaans as a language:

> Unlike many other language movements, the attempt, starting in the 1870s, to develop Afrikaans as a public language was not the work of the higher stratum of a group who used it as a tool to advance their career chances. Its initial impetus came on a wave of religious enthusiasm that sprang from the Revival of the 1860s and, more particularly, the desire to disseminate the Christian message to the very poor, both white and coloured. A simple medium was required to spread the Christian message more widely in print.
>
> (p. 216)

As issues of school attendance and medium of instruction became more concerning, the Education Department took steps to be more accommodating of Dutch. In 1891, the Education Commission was mandated to consider, among other issues, instruction in both English and Dutch. The Commission Report was sympathetic to the submission by the *NGK* Synod and *Taal* Congress on matters of language of instruction, the inclusion of the Dutch language in schools and teacher qualifications, and the inspectorate. In 1882, the use of Dutch was permitted as a medium of instruction and a school subject, though the system remained English-oriented and its organisation meant that instruction in Dutch would be enabled but not championed by the Department. As it was, in more remote areas, Dutch was increasingly giving way to Afrikaans as the language of daily use.

The colonial government's modernising intent was to establish a system of schooling that looked to Britain for design and to fellow British colonies such as New Zealand and Australia for comparisons. The annual SGE Reports show its governmental telos: to prepare young white people for future opportunities in the expanding colony with its modernising economy. In contrast to the modernist orientation of the government, the major concern of the *NGK* and of mission schools was pastoral: Christian formation and saving souls. For missionaries, the telos was conversion to Christianity in its westernised forms and an emphasis on the virtues of temperance and work. For the *NGK*, pastoral care was in part inspired by its own revivalist movement and by concerns about the fate of the growing number of poor white children (concerns not paralleled for poor black children). Certainly, in the town of Carnarvon, the foundations for schooling were laid by the Rhenish missionaries, but it is also the case that the *NGK*'s efforts were decisive in shaping the state system for the benefit of white children.

The sections above suggest the different affordances of schooling in relation to population dispersal, social class, and racial designation. Official education policies presented the template for a system that favoured the interests of white people with financial resources living in towns; this was not a system that was able to reach the majority of the population in the period under consideration. Neither in its intended design nor in its actual reach did it match the livelihoods and meanings of the majority of white Afrikaner farmers in this period, and it made no attempt, beyond subsidies to mission schools, to cater to black people. For the former, the *NGK* was a major sponsor; for the latter, the different missionary societies. In other words, in places beyond the reach of modernist governmental power, pastoral power prevailed. The pastoral power of the *NGK* was increasingly focused on the formation of its 'own' people, fusing religious formation with an emergent Afrikaner nationalism that was racially exclusive. Mission schools, receiving state subsidy at lower levels than public schools, were under-served throughout this period as they strove to build and sustain Christian and westernised communities among black populations, and most provided little more than basic education.

Setting the trajectory for schooling in Carnarvon

Against this backdrop analysis of the different actors and activities making up schooling in Carnarvon, I return now to look at how these different strands interlaced to form a distinctive palimpsest for future schooling in this local place. As mentioned earlier, the story of schooling in Carnarvon begins with the Rhenish mission school which Alheit had opened in 1848. This school served a growing and stable community, with teaching led by German-educated missionaries and at times a designated teacher from Germany. For many years, white children attended the school, alongside Xhosa and Baster children. From 1856, the school received aid from the Cape government as a second-class mission school, and its staff was expanded to include trained Coloured assistants.

The SGE Report of 1858 provides a detailed account of the school, its attendees, and the subjects it offered, which is worth repeating in full to illustrate the nature of reporting in this period as well as the conditions of schooling:

> **Rhenish Mission School, Schietfontein**
>
> In the main settlement under the Kareebergen, to which this school belongs, there is a population of 1,500 souls, of whom 100 belong to the families of emancipated slaves, and 1,400 to those of other persons of color, chiefly bastards and Kafirs. It is situated in the division of Beaufort West, and on what was the boundary of the colony, previous to the proclamation of 1848.
>
> In the day-school of this institution there were enrolled in 1858, 85 boys and 112 girls; total, 197. Of these, 13 were European, 21 the children of emancipated slaves, 73 of Kafirs, and 90 of other persons of color, chiefly bastards. The greatest number on the roll at any one time was 162, the least, 108, the average, 135, and the average daily attendance, 110. The number admitted was 108, transferred from the roll of the former year, 89, and withdrawn, 51. Twenty-five of the pupils enrolled were under 5 years of age, 101 between 5 and 10, 63 between 10 and 15, and 8 above 15.

The number of days the school was open, exclusive of Sundays, was 302, and on each day for 6 hours, namely, 4 in the forenoon and 2 in the afternoon. Of the 197 enrolled, 54 were present two thirds of the school days and upwards, 73 from one half to two thirds, and 70 under one half.

In the reading classes, there were 43 in the first, 18 in the second, and 18 in the third or lowest. The following is a statement of the subjects taught the reading classes, and the hours weekly assigned to each:

Bible and church history, 6 hours.
General history, 2 ditto.
Elementary geometry, 2 ditto.
Arithmetic, 4 ditto.
English and Dutch grammar combined, 2 ditto.
Exercises in translation, 4 ditto.
Geography, from the maps, 2 ditto.
Drawing (occasionally).
Sacred music, 2 ditto.
The remaining hours given to exercises in reading and writing.

In the non-reading classes, there were 17 that could spell correctly and read easy sentences, 36 that can only spell, and 35 in the alphabet.

The total number of pupils in the reading classes was 109, or 11 in every 20 enrolled. In the non-reading classes, the number was 88, or 5 in every 11 enrolled.

Of the 51 withdrawn, 26 were from the reading classes, and 25 from the alphabet and spelling classes. The reason assigned for withdrawal was, the parents having to leave the institution on account of their live-stock, and unable to leave their children behind. From the nature of the country, which is entirely pastoral, it is necessary to change the pastures at certain seasons of the year, which leads to the continual withdrawing of the children, which greatly impedes their progress and weakens the efficiency of the school.

The school is conducted by a trained teacher from Germany, who is aided in conducting the more elementary classes by four female assistants. The languages used in the school are English, Dutch, and Kafir.

In the Sunday school of this mission there was an aggregate attendance of 200 pupils, and in the evening of 110; of the latter, 70 were adults and 40 children. The number of adults on the station that can read is 250, of children, 109.

The resident missionary, in his general report, states that the colored families in that part of the country who do not avail themselves of the means of instruction at the missionary institution are greatly neglected; and there were very few instances in which natives receive any instruction among the farmers. Some of the children of the neighbouring farmers are now sent to Victoria West to school. Other farmers keep private teachers for short periods. Generally speaking, education in this district is of a very low standard.

Statement of income and expenditure: Income – Government grant, £30; voluntary contributions, £10 10s; school pence, £4 10s; total, £45. Expenditure – Teacher's salary, £45.

(SGE Report, 1858, pp. 130–131)

It is interesting to note that at this point (1858) when the Rhenish Mission School was the only school in the area, the attendance of a small number of white children is recorded in what was quite a flourishing school by the standards of the day. It is also interesting to note the details that concerned the Education Department in its school inspection: racial classification of attendees (with gender a secondary concern); the complex calculations of actual attendance in contrast to enrolments (a problem that bedevilled statistical records at the time); the main division between 'reading classes' and 'alphabet and spelling classes' and the numbers at each level; the detail recorded for the time allocation of different subjects; and brief comments on the teachers and their qualifications and on the general conditions of life in the area. Also noteworthy is that the government grant at that time did not cover the recorded salary of the teacher.

A few years after this, in 1864, a new and larger mission school was built at the settlement, and by 1891, when the tenure of the second missionary Reverend Sterrenberg ended, the school had extended its classes to offer four years of instruction. Although white children ceased to attend the school once other schools were established, enrolments remained fairly constant at 120–130 students – a higher number than those attending the white school.

It is clear that the schooling established by the Rhenish missionaries in Carnarvon provided a continuing presence in the community well into later years. The school suffered disruption during the South African war (1899–1902) when buildings were taken over by a British garrison and the resident missionary was briefly interred. After the war, the school was returned to the Rhenish community and restored to operating condition. Two new classrooms were built, and numbers continued to grow – though the school was exclusively for Coloured and Xhosa children and offered only primary classes. When the missionary Hartwig left in 1943, it had 360 students and 10 teachers.

Strassberger provides the following statistical picture of the Rhenish Mission School for the period from 1850 to 1930 (Table 4.2).

There is no doubt that the mission school was a strong part of the institutional fabric of the church during the years of its operation, and that it provided education of sufficiently good standard to enable a number of its (primary school) graduates to move to secondary schools elsewhere.

Looking ahead from this time, the mission school clearly provided the basis for future primary schooling for Coloured people in the area. In 1943, the RMS, which had always worked for the self-sufficiency of its mission

Table 4.2 Rhenish Mission, Schietfontein/Carnarvon

Date	*Church attendees*	*Full church members*	*School pupils*
1850	191 baptised	88	140
1874	1200	288	211
1902	1868	596	202
1910	2145	855	176
1930	2503	1027	358

Source: Reconstructed from Strassberger (1969, p. 80, 86–89).

stations, found itself no longer able to maintain its settlements in South Africa. It negotiated with the Carnarvon Rhenish church community and the *NGK* for the church and its school to be taken over by the *NGK* as part of the *Sendingkerk* – in effect as an Afrikaans Coloured missionary church and school.[18] The school continued its steady growth as a segregated mission school, with 500 students and 13 teachers in 1947.

After the electoral victory of the National Party in 1948, apartheid legislation intensified the patterns of segregated schooling, progressively moving Coloured mission schools under the control of provinces. In 1963, the Coloured Persons Education Act moved the control schooling for Coloureds from provinces to a national Department of Coloured Affairs, cementing their separation from white education departments. In the 1960s, after a fire, the Carnarvon school was relocated across town and renamed *Laerskool* Carel van Zyl after a local white dignitary – and this has remained the official name of Carnarvon's current primary school. In 1984, a secondary school was built in the Coloured township of Bonteheuwel across the dry riverbed from the town – providing post-primary education for the first time to Carnarvon's Coloured residents. It is this school building that was allocated as the site of the town's primary school in the restructuring that took place in 2007, as will be explained later. What this brief synopsis suggests, in a Lefebvrean sense, is that the early schooling practices in the town established a particular pattern whose resonance has continued, albeit altered in its interaction with other movements and flows in this space.

The education of white children followed a different trajectory. Once an official *NGK* community was formed in Carnarvon, with a grand church being built in 1874, it was the activities of the *NGK* and the Cape Department of Education that shaped white schooling the town. In 1875, a Second Class Public School was opened, but it did not have an auspicious start, with the first teacher being dismissed for misconduct.[19] Disputes arose about the lack of religious instruction, and there was also some contestation around funds. Moreover, in Kitshoff's (1974) words, the school had *'n Engelse gees* (an English spirit), which was not acceptable to the white community and its church (p. 137). The SGE Report of 1882 noted that at this point, of the 312 children in the district, only 49 white children were 'receiving fair instruction' and another 96 'receiving inferior instruction', presumably referring to farm schools. It noted that the RMS school was flourishing, with four teachers and 167 students.

In 1884, a free private *NGK* school was opened by the pastor of the local *NGK*, who was related to the Paarl-based Reverend van der Lingen mentioned earlier, and who identified himself with the early movement for Christian-National Education. The intention in setting up this school was to give more emphasis to the teaching of Dutch and religious instruction. Kitshoff (1974) notes, however, that when this school began to take increasing numbers of poor white children, wealthier parents in the community started sending their children to the state-aided public school instead.

The SGE's Report for 1895 noted that more than half of the town's children of school-going age were not receiving any instruction, while more than two thirds of those on farms received no instruction. The Report also noted that a number of poor children could not afford fees, though more than half of them did attend school.[20]

Kitshoff (1974) notes that in 1895, the *NGK* negotiated an amalgamation of the free school with the state school, on the understanding that *Hollands* (most likely in local Cape variant) would be taught, and a place would be made for religious education in the curriculum. To ensure that these principles were adhered to, four members of the Church Council would serve on the School Committee. Implicit in these arrangements was the racial segregation of the school, since the *NGK* was a whites-only church by that stage, with racially separate mission churches serving black communities. The combined public-with-poor school grew in size, and in 1896, had a new school building with six classrooms. However, a continuing issue for the *NGK* as well as the Education Department was how to provide for children on farms, and added to this was an increasing need to assist poor white children in the town, whose numbers were growing during and after the 1890s.

From 1905 onwards, with the School Board Act, compulsory schooling was introduced for white children in the Cape.[21] In 1912, after the Union of South Africa was formed, the Cape Education Department opened a hostel at the school in Carnarvon to cater for white farm children, and in 1915 the church opened its own hostels, which it supplied with '*koshuisouers*' ('hostel parents'). When schooling became compulsory for all white children in 1917, the Cape government subsidised the hostel accommodation for poor white children, as the poor white crisis deepened.[22]

Kitshoff (1974) views the hostel system together with free and compulsory schooling as launching a new era for education in Carnarvon. He notes that it brought a greater concentration of children to the school in the town (up to 450 in 1922), attracting children from nearby towns. In 1921, the school was extended into a high school, and continued to expand. It was this school, later known as *Hoërskool* Carnarvon, that bitterly resisted change after 1994, holding on to its sense of past traditions. However, Kitshoff's comments about a new era for education certainly did not apply to all schools in Carnarvon, where segregation and unequal provision was consolidated.

Summing up on schooling

To conclude this chapter, a number of points about schooling in Carnarvon are worth considering.

First, it is interesting to note that public schooling had been racially segregated in Carnarvon since its establishment there – though prior to this the mission school had admitted white children alongside Xhosa and Baster. From the start, public schooling in the town was strongly connected to the *NGK*, which took steps to ensure that the Dutch language or its emerging Afrikaans variant was taught, and that its particular approach to Christianity was fostered in the school. Alongside and supplementing the state's efforts, the *NGK* provided free schooling for poor whites, as well as feeding schemes, and it provided hostel facilities where care was taken to provide *koshuisouers* to look after school children. In contrast to the nondenominational if not secular schools envisaged by the Cape Education Department, *NGK* schools provided a pastoral context that fostered a particular Christian and language/cultural identity, even before the Christian-National Education movement took root in a later period.

While it might be an overstatement to say that the formal curriculum requirements were less important to this white community than its cultural and pastoral nurturance, it is certainly the case that many white children in towns as well as on farms had little or no formal instruction during this period. Child labour was extensive, and farm parents were often indifferent to schooling. Interesting also is evidence of a social class differential in schooling preferences as poor white numbers grew in the town. The 'childhood' discussed by Duff's (2015) study of the south-western Cape Colony was not the experience of most children in the more remote and sparse regions of the Karoo, where many children were poor, worked, and did not attend school.

In terms of the Rhenish mission, instruction in its earlier years was given by German pastors or specialist teachers from Germany, and Strassberger mentions the strong Christian moral code and work ethic that the church promoted. Although many Xhosa and Baster children living around the town did not attend school at all (and it is likely that attending school would have been a marker of relative privilege), the mission church and later the *NGK Sendingkerk* was active in the social life of the town. Louise Boezak, current resident, political stalwart, and pillar of the church, commented that this schooling provided her with an educational base to further her secondary schooling at a convent in Cradock, and to study further for post-school qualifications.[23] It provided her father, Andries Boezak, with a sufficiently good education to write music and lead the church brass band in the town as well as to work as clerk and translator for the magistrate of Carnarvon.

The continuing presence of the mission as well as its association with the schooling of the Coloured community certainly left its mark on the town. Names of the major streets reflect missionary presence (*Alheitsraat, Sterrenbergstraat, Biesenbachstraat*), but more important indicators are numerous church-based activities. In the 1980s, the *Sendingkerk* in Carnarvon was associated with the international Reformed Church movement, which declared apartheid to be a heresy – as discussed in more detail in a later chapter. Nor should it be forgotten that the very existence of the town as a place for Xhosa and Basters to have legal land ownership rights was due, to some extent at least, to the efforts of missionary Alheit. I return to the story of the *Sendingkerk* in later chapters, showing its importance in the anti-apartheid movement in the 1980s and in bringing community members together to engage with the SKA.

Notes

1 Though mention is made here of childhood, there is no attempt to address the literature on colonial childhoods, or what Rebecca Swartz (2019) identifies as the 'intimate affective ties between adults and children' (p. 10) and the 'emotional frontiers' of childhood (p. 15).
2 In focusing on the provision of schooling, this account is necessarily partial. It does not look at classroom practices, or the experiences of those involved – teachers, students, parents, and communities. This would require historical research beyond the scope of this study.
3 Strassberger (1969, p. 79) notes that this early school had 130 students, including both adults and children. Strassberger's text is the source for details on the RMS station given here. Ross (2018) gives further insights into life on mission stations. There are extensive debates on missionaries and mission schooling in South Africa, which are not referred to here, where the focus is on the RMS in Carnarvon.
4 The following section draws its information from Blue-Books, SGE Reports to the Cape Parliament, and Commission of Inquiry Reports – all primary sources – supplemented by the magisterial text of E. G. Malherbe (1925). (The term bluebook is used broadly for parliamentary papers. Annual Cape of Good Hope Blue Books were published from 1821 to 1885, after which time a Statistical Register was published. SGE Reports were published annually in parliamentary annexures and appendices.) The education system changed as the Cape economy grew and diversified and government competence expanded (to become Representative Government in 1854 and Responsible Government in 1872). It also needs to be said that there is a large literature covering economy, politics, and social relations in the Cape Colony, which this chapter draws on generally but does not cite.
5 In 1883, there were 67 such schools, and in 1887, there were 179. A system of community-funded boarding hostels operated alongside public schools as well.
6 Malherbe (1925, p. 102). Soon after this, the Destitute Children's Relief Act was passed in 1895.
7 SGE Dale was a firm believer in the importance of 'voluntarism', meaning parental support for and contribution to schooling. His views on this matter are clearly set out in numerous Annual Reports.
8 See Malherbe (1925, p. 92) for numbers: In 1840 there were 4000 students; in 1860, double that; in 1880, 72,281; in 1890, 104,000.
9 See especially the Ross Report of 1883, the Commissions of 1879 and 1891, and the Special Reports of 1901.
10 The Ross Report is marked as 'provisional', since Ross became ill and died before it was formally adopted.
11 The administration was reorganized by SGE Muir, appointed in 1892. It was Muir who introduced aid for poor schools.

12 Malherbe provides the following comment:

> The Commission found that of the 99,280 white children of school age (5–14 years) in the Colony, only 41,037 were enrolled in aided or private schools or taught at home by governesses, leaving 58,243 white children unaccounted for. These facts were decidedly at variance with Dale's reports.
>
> (1925: 125)

13 Malherbe's (1925) comment on sources is salutary:

> The gathering of statistics on the early periods of education has also been very difficult. Owing to glaring discrepancies in the different records of the same facts the accuracy of the many of the figures during the early stages is, therefore, extremely doubtful ... Then, too, the inadequacy of the Education Department Reports and their lack of uniformity at the hands of different Superintendents even in the same Province, has made the work exceedingly difficult and necessitated many gaps.
>
> (p. 13)

14 There is an extensive literature on this, which lies beyond the scope of this study.
15 Duff's (2015) research provides an excellent account of this in regard to the south-western Cape, but there is no comparable study of the Cape Colony as a whole.
16 The Destitute Children Relief Act of 1895 highlighted the needs of white children in poverty. These concerns about poor whiteism escalated in the early 1900s and are set out in the Carnegie Commission of Enquiry into Poverty in 1932.
17 See Giliomee and Mbenga (2007) for a description of this, including a picture of the first written Afrikaans, in Arabic, in a Muslim prayer text (p. 71).
18 See records of negotiations – *NGK Argief.* This legacy-Rhenish mission church, now the *Vereenigde Gereformeerde Kerk* still draws a solid congregation. Up until the end of this research, Reform congregations remain segregated in two churches in the town.
19 Report of the SGE of Education for the year ended 30 June 1882 (published 1883).
20 Report of the SGE of Education for the year of 1895 (published 1895).
21 This was done gradually and under the decentralized School Board system. By 1917 school was compulsory to age 16 or Std VI.
22 The government provided an amount of £19 per child per year. In 1923, there were 170 hostels run by the *NGK.*
23 Interview, Carnarvon, November 2015.

References

Official publications

Blue-Book for the Colony of the Cape of Good Hope. (Annual). Cape Town: Saul Solomon.

Report of the Superintendent-General of Education. (Annual). Cape Town: Saul Solomon; Cape Town: Government Printers. (Years cited: 1858, 1887, 189, 1895.)

Report of a Commission Appointed in Accordance with a Resolution of the House of Assembly, to Enquire into and Report Upon the Working of the Education Acts in Force in This Colony, 1879 (1880) Cape Town: Saul Solomon. (de Villiers Report).

Special Reports on the Systems of Education in Cape Colony and Natal. (1901). London: H.M.S.O.

Cape Colony: Annexures to the Votes and Proceedings of the House of Assembly

G3–92 Reports of the Commission appointed to enquire into and report upon certain matters connected with the Educational System of the Colony 1891. (Barry Report)

G12–83 Preliminary Report on the State of Education in the Colony of the Cape of Good Hope, 1883. (Ross Report)

G7–59 Report of an Inquiry into the Claims of Certain Natives Residing at the Missionary Institutions of Amandelboom and Schietfontein, in the Division of Beaufort, to the lands on which they are located. Presented to both Houses of Parliament by command of His Excellency the Governor, 1859 (Auret Report).

Secondary sources

Bade, K. J. (1975). Colonial missions and imperialism: The background to the fiasco of the Rhenish Mission in New Guinea. *Australian Journal of Politics & History, 21*(2), 73–94.

Bundy, C. (1984). *Vagabond Hollanders and runaway Englishmen, white poverty in the Cape before poor whiteism* (pp. 11–23). Collected Seminar Papers. Institute of Commonwealth Studies, 33. ISSN 0076–0773.

Duff, S. E. (2015). *Changing childhoods in the Cape Colony: Dutch Reformed Church evangelicalism and colonial childhood, 1860–1895*. London: Palgrave Macmillan.

Etherington, N. (2005). *Missions and empire*. Oxford: Oxford University Press.

Giliomee, H. (2011). *The Afrikaners: Biography of a people*. London: Hurst & Company.

Giliomee, H. & Mbenga, B. (2007). *New history of South Africa*. Cape Town: Tafelberg.

Kitshoff, M. C. (1974). *Kudde van Carnarvon: Nederduitse Gereformeerde Kerk, 1874–1974*. Carnarvon: N.G. Kerk.

Lefebvre, H. (1991). *The production of space*. Oxford: Blackwell.

Mackenzie, C. G. (1993). Demythologising the missionaries: A reassessment of the functions and relationships of Christian missionary education under colonialism. *Comparative Education, 29*(1), 45–66.

Malherbe, E. G. (1925). *Education in South Africa*, Vol. 1. Cape Town: Juta.

Mbembe, A. (2008). What is postcolonial thinking? An interview with Achille Mbembe. *Eurozine*, 9 January. https://www.eurozine.com/what-is-postcolonial-thinking/ Accessed 15 November 2019.

Newton-King, S. (1999). *Masters and servants on the Cape Eastern Frontier, 1760–1803*. Cambridge: Cambridge University Press.

Ross, R. (1993). *Beyond the pale: Essays on the history of colonial South Africa*. Middletown, CT.: Wesleyan University Press.

Ross, R. (2018). The archaeology of missions: Afterword. *Journal of Southern African Studies, 44*(4), 743–747.

Strassberger, E. (1969). *The Rhenish Missionary Society in South Africa, 1830–1950*. Cape Town: Struik.

Swartz, R. (2019). *Education and empire: Children, race and humanitarianism in the British Settler Colonies, 183–1880*. Cham: Palgrave Macmillan US.

van Sittert, L. (2016a). Children for ewes: Child indenture in the post-emancipation Great Karoo: c. 1856–1909. *Journal of Southern African Studies, 42*(4), 743–762.

van Sittert, L. (2016b). Working children: Rural child labor markets in the post-emancipation Great Karoo, South Africa, 1856–1913. *Journal of Family History, 41*(1), 39–64.

5 Apartheid's local forms

Municipality, school, and church in Carnarvon

In focusing on two time periods – the establishment of the town of Carnarvon in the mid-1800s and the pivotal moment of change after the end of apartheid – this study skips over highly significant events in South Africa's history. More than a century lies between these two periods of time, and this was a century of highly significant change for South Africa. Most notable are the economic and social transformations of the mining revolution sparked by the discovery of diamonds (1867) and then gold (1886), accelerating capitalist development, urbanisation, and the growth of manufacturing. Politically most notable was the imposition of apartheid rule under the National Party (1948–1994), the drafting of the Freedom Charter by the Congress Movement in 1955, and the growing militancy of the African National Congress (ANC), Pan Africanist Congress (PAC), and other opposition groups banned by the government in the 1960s.

It is simply not possible to do more than provide a sketch of changes during this period and how they impacted on schooling in this place. And, much as it would be relevant to explore this period in relation to debates on colonialism and coloniality, it is simply not possible to provide an analysis of the apartheid state formation within the pages of this chapter, beyond a few brief comments. Nor is it possible to elaborate on the activities of political resistance movements, significant though they were in bringing about the demise of apartheid and colonialism. Rather than attempt to provide a narrative overview, this chapter merely sets out the architecture of apartheid legislation, without detailing its particular racial and capitalist form, spatial segregation, control of labour, and everyday humiliations. Instead, the chapter picks up the mantra repeated by several of the white townspeople of Carnarvon during fieldwork for this study: '*First they took our municipality, then they took our school, and now they want our church*'. This mantra speaks to a sense of possession and dispossession that has powerful historical roots that stretch from the past to the present. It also provides a lens to focus on what apartheid looked like in three aspects of social life in this town: local government, church, and schools.

Apartheid's brutal laws

As set out in Chapter 1, the apartheid state may be understood as a form of racial capitalism. Apartheid's early architecture included a set of laws of overt racial brutality, related also to measures of labour control and spatial restriction. It is important to recognise that these laws were promulgated on top of existing segregationist practices and legislation, while at the same time not conflating apartheid with prior practices. Most notable continuities were the Land Acts of 1913 and 1936, which allocated around 80% of the land to the white minority; 'job reservation' legislation, which protected white labour from black competition; measures to control the movements of black people; and legislation that removed the last vestiges of political voice from Coloured people.

The following outline of apartheid legislation shows its reach into all aspects of public life:

- *Prohibition of Mixed Marriages Act* (1949) declared marriages between whites and other designated racial groups illegal;
- *Immorality Act* (1950) banned sexual relations between whites and blacks;
- *Population Registration Act* (1950) classified people into rigidly bounded racial classification groups;
- *Group Areas Act* (1950) stipulated spatial segregation on racial lines and enabled forced removals of people;
- *Bantu Authorities Act* (1951) locked the majority of the population into fragmented and cramped 'tribal reserves' that had their origins in the 1913 Land Act and colonial conquest before that;
- *Native Laws Amendment* Act (1952) narrowed the categories of black people who could live in cities and extended 'influx control' (placing limits on black urbanisation);
- *Natives (Abolition of Passes and Coordination of Documents) Act* (1952), while ironically named, constituted the 'pass laws', and allowed for the removal of black people from cities with 72 hours' notice;
- *Reservation of Separate Amenities Act* (1953) stipulated the segregation of amenities and stated that separate amenities need not be of equal quality – the Act that resulted in 'Whites/Europeans Only' and 'Non-whites/Non-Europeans' signs in public places;
- *Separate Representation of Voters Amendment Act* (1956) removed Coloured people from the common voters' roll;
- *Bantu Labour Relations Regulation Act/Natives Settlement of Disputes Act* (1953) prohibited black workers from joining registered trade unions and prohibited their strike action;
- *Industrial Conciliation Act* (1956) extended job reservation to protect white workers; and
- A raft of security legislation prohibited all forms of protest against the government, including the *Suppression of Communism Act* (1950), the *Public Safety Act* (1953), and the *Official Secrets Act* (1956).

In education, the raft of laws that undergirded segregated and unequal provision along the lines of population registration were set up in the 1950s and 1960s. They included:

- *Bantu Education Act* (1953), the infamous act which closed mission schools and brought black education under state control;
- *Coloured Persons Education Act* (1963), which formalised the segregation of Coloured schools under a separate department;
- *Indian Education Act* (1965), another segregating measure;
- *National Education Policy Act* (1967), which enshrined Christian National Education for people classified as white, defining both of these concepts in line with Afrikaner nationalist traditions and requiring the strict separation of English- and Afrikaans-speakers in white schools;
- *Extension of University Education Act* (1959), another ironically named Act, which established the principle of segregated and separate institutions at the university level.

Much as apartheid might appear as a 'grand design', particularly when its architecture is presented in this way, it is more accurate to regard it as roughly crafted and adapted along the way, as Deborah Posel (1997) articulates with particular clarity in her analysis of the apartheid state:

> Whatever its superficial appeal, … the notion of a single master plan fundamentally misrepresents the political processes whereby Apartheid was built, greatly exaggerating the extent of the continuity, control, and long-term planning involved. The source of these distortions is a limited, often crude, analysis of the workings of the state. The notion of a 'grand plan' suppresses interest in the internal workings of state departments, because policy decisions and administrative practices seem merely to have been transcribed from a ready-made blueprint with little further ado … [A]n alternative picture is painted of the making of Apartheid, as having been forged through a series of struggles within and beyond the state, which forced the architects of state policy to adapt and revise many of their original strategies. Uncertainties, conflicts, failures, and deviations, although often less visible than the continuities and triumphs of Apartheid, were fundamental to its development.
>
> (p. 4)

Before looking at these measures in the context of Carnarvon, it is important to add another set of moves taken by the apartheid state to intensify the spatial division of land while tightening its control over populations. First, through the Bantu Authorities Act, the state turned the 1913 and 1936 reserve lands, supplemented with other land purchases, into separate 'Homelands' to which black South Africans were designated on the basis of (presumed or actual) ethnic origins. In a variant of the indirect rule characteristic of colonialism in Africa, these 'Bantustans' were administered by chiefs and officials paid by the South African government, and were steered towards 'independence' through increasing powers of self-government. Separate administrations, state services, and infrastructure were established, including separate departments of education for each Bantustan. This policy of spatial division deprived African people of their South African citizenship, and more than three and a half million people were forcibly removed from urban areas and other white-designated spaces to Bantustans. (As was mentioned in Chapter 1, the establishment of Bantustans and their 'independence' further complicates the analysis of South Africa in terms of theories of colonialism).[1]

It is interesting to note the ideological justification given for separate homelands by the then Minister of Native Affairs, Dr Hendrik Verwoerd, speaking in 1950:

> In my soul I believe that the apartheid policy is the only policy which can bring freedom because it has the interests of both groups (white and black) at heart. As his guardian, we must see that the Native is kept by means of apartheid, to the greatest extent, and if possible to an increased extent, in territories where he can live his own life. We would also like to see him anchored in the language and customs of his own people so that he cannot become an imitation White. The Afrikaner is looking for a way by which a person can live and let live. To the Native we say: let your ambition lie in the building up of your own people. We desire for them ideas and a future such as those we want for ourselves, but their future and ours are separate.
>
> (Quoted in SAIRR, 1950, p. 3)

What Verwoerd neglects to include in this endorsement of 'live and let live' are the economic exploitation, political oppression, violent coercion, and racial denigration on which apartheid depended.

Estimates are that about 44% of the country's population (17 million people) were living in Bantustans by 1990, with about 800 traditional leaders given financial support by the South African government (Oomen, 2005). Despite the formal dissolution of Bantustans in the political changes of 1994 and the restoration of South African citizenship to those living in Bantustans, the patterns of land holding, patriarchy, and traditionalism as well as the powers associated with Bantustan administration have proven extremely hard, if not impossible, to shift – a point I return to in the next chapter.

The town of Carnarvon, located in the Karoo heartland, was never part of any reserve or Bantustan, hence this study makes only passing reference to what is a significant issue for a broader study of decoloniality in South Africa – the continuation of traditional law and land ownership under current constitutional arrangements.

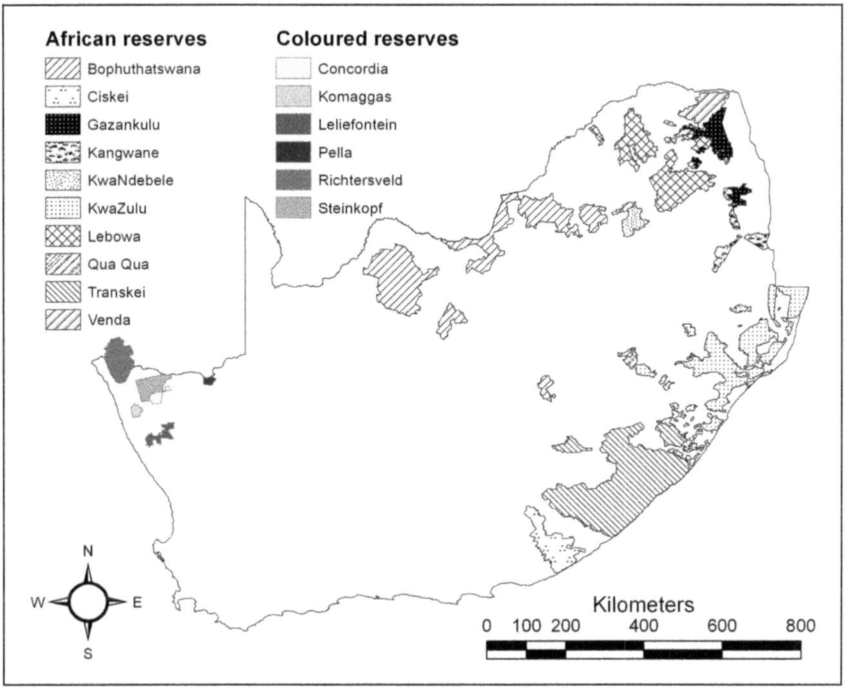

Map 5.1 Bantustans of South Africa.
Source: Copyright © Timm Hoffman, 2014. Reproduced by permission of M. Timm Hoffman and Sam Jack.

The next section uses the mantra of municipality, church, and school to show how apartheid affected the lives of people in the town of Carnarvon. When viewed from the vantage point of this small and marginal place, these pictures cannot capture the scope of apartheid measures which impacted differently upon daily lives in different parts of South Africa. Nonetheless, this vantage point provides the opportunity for closer examination of the texture of social life under apartheid.

'The municipality' and local government under apartheid

In the town of Carnarvon, the apartheid mantle fitted easily over existing deep inequalities. Generations of white ownership of farms and resources erased the history of earlier ownership and even earlier occupation of the land by others, and supported a sense of justified Afrikaner emotional attachment to the land and the institutions of the town: '*its* municipality', '*its* school', and '*its NG* church'. Residential areas in the town had consolidated on predominantly racial lines, with the majority of the original inhabitants residing in small dwellings in De Bult and the around the Rhenish Mission

Church. The dissolution of the outer and inner commonage together with the 1913 Land Act meant that there were no possibilities of land ownership for farming activities by people of colour.

Later, in the apartheid logic of segregation, a new township, Bonteheuwel, was built for Coloured people on the other side of the town's dry riverbed. In 1956, the Group Areas Act made its first assault on the town, moving to displace a number of families from an area designated 'white', one of these being the Boezak (also spelled Boesak) family. This was staved off for several years by the Boezaks who obtained special permission to stay in their established suburban home. In 1965, Louise Boezak moved with her mother and sister from their spacious house to a much smaller house in one of the town's main roads near the mission church and De Bult.[2] Louise Boezak recalls that the meagre compensation paid for the house was divided among 12 surviving children, with her brother Alwyn deciding to frame his £12 share as a constant reminder of this injustice – a cheque for R1,103.[3]

An attempt to demolish and relocate De Bult – the small, long-established Coloured area near the former *Sendingkerk* – did not come to much, partly because the white community had no real need or appetite for that part of town (which was near the cemetery), and there was no alternative housing for its relatively few residents. Louise Boezak fought against this proposed demolition in court, and ultimately the municipality did not proceed, though it won the case. De Bult remains a part of the town currently known architecturally for its typical Karoo houses, though many of these houses are in disrepair.

In 1972, the Department of Community Development proclaimed Schietfontein to be a white area, but no further steps were taken to change occupancy. A putative expropriation order in 1981 was met with resistance, including from a section of the white community, and was withdrawn in 1983. Schietfontein remains as a small and impoverished Coloured residential area just beyond the outskirts of the town, and its current residents have lodged a land claim for ownership to be recognised. Perhaps most shocking of all was the attempt to remove the *Sendingkerk* from its historic 1858 buildings – deemed by the municipality to be in a 'white' area of the town – and turn these into a museum. The matter was taken to court, and the *Sendingkerk* was given a special permit to occupy its church for religious purposes.[4] These assaults by the Group Areas legislation could be seen as another wave in the rhythms of possession and dispossession, which, even if small in scale, had historical resonance of serious weight.

Before apartheid ended in 1994, white townspeople assumed that their interests should prevail in the running of the town, and that as ratepayers they should control the municipality. In this mindset, 'others' were entitled to remain in the town in subordinate positions only, living in separate spaces, contributing their labour to farms and businesses owned by whites, and served by separate (and poorer) churches and schools. Later, a new and different 'management committee' was set up to advise the municipality of the township's interests as part of apartheid's local government system.

Carnarvon's municipal arrangements need to be understood within the context of apartheid's broader set of local government arrangements for the country as a whole. In some ways, the relative isolation of the town insulated its white community from the crises that were growing across the country at the local government level in the period before the changes of 1994. At the local level, the apparent coherence of apartheid's architecture had in fact covered over a messy patchwork of arrangements, all embedding a racial geography of segregation and inequality. In brief, the provision of services at the local level across the country was racially differentiated, unequally funded, and organisationally mixed (see Cameron, 1993; de Visser, 2009; Picard & Mogale, 2015). Cities and towns like Carnarvon had municipalities funded by local rates and taxes to deliver services to white communities; Coloured and Indian 'management committees' were introduced in some places to act as advisory bodies to white municipalities; and responsibility for African townships fell under separate Administration Boards. There was almost no provision made for the majority of people living in rural areas, and there was very limited local government in Bantustans, where state-funded 'traditional leaders' were given powers over land and development. In effect, this mixture of arrangements protected the revenue base of white municipalities by limiting their responsibilities for providing services to other racial groups. At the same time, no adequate revenue base was provided for other local structures to serve the needs of the majority of the population. In addition, the multiple indignities of 'petty apartheid' confronted people in their daily lives wherever they lived.[5]

By the early 1980s, these messy and inadequate arrangements were in crisis. In response to growing political protests, the apartheid government introduced new constitutional provisions for national and local government. Through constitutional changes, a Tricameral Parliament was established in 1983 to provide representation for people classified as Coloured (in the House of Representatives) and Indian (in the House of Delegates), with the white House of Assembly as the main seat of national government. To bring some order to the uncoordinated – if not chaotic – arrangements for service provision at the local government level, a system of Black Local Authorities and Regional Services Councils was introduced. However, far from propping up apartheid, these arrangements served as flashpoints for community mobilisation against what were seen as illegitimate structures. In 1983, the United Democratic Front (UDF) was launched as a major popular movement against apartheid, organising alongside the growing democratic trade union movement (the Federation of South African Trade Unions, later the Congress of South African Trade Unions). Protests in education, starting with the June 1976 Soweto protests, were an ongoing and very significant part of the broader protest movements, as will be discussed later.

The burgeoning protests that were organised through civic movements and other community organisations against the unacceptable conditions at the local level led to the virtual collapse of the local government system. The confused nature of local government – its mix of structures and particularly its funding distortions – meant that major structural changes were required. As a result, a separate Local Government Negotiating Forum was established alongside broader constitutional negotiations in the transition period of the early 1990s, and the Local Government Transition Act (1993) set out a phased approach to local government changes. I pick up this theme in the next chapter.

'The church'

As mentioned earlier, when the Rhenish Missionary Society closed down its activities in South Africa in 1943, negotiations were undertaken for its Carnarvon church to join the *NGK* as a Coloured mission church, since both shared a Reform Christian tradition. The incorporation of the *Sendingkerk* into the *NGK* cemented the segregation of the churches in the town, and it also signalled that there would be a substantial differential in wealth between the two churches, since *NGK* congregations held their finances separately. These two separate reform churches still operate around the corner from each other, serving segregated (and financially unequal) congregations.[6]

During the 1980s, many South African church leaders joined anti-apartheid movements, with the South African Council of Churches (SACC) being a focal point for protests. In this context, it is not surprising – but nonetheless highly ironic – that the *NGK* was challenged from within by its own segregated church structures.[7] In the 1980s, members of the Coloured and African branches of the *NGK* took the decision that their segregated status should end, and they began to work together, finally forming the Uniting Reformed Church in Southern Africa (URCSA) in 1994. Meanwhile, in 1981, a grouping of Dutch Reformed, Presbyterian and Congregationalist theologians had formed the influential Alliance of Black Reformed Christians in South Africa (ABRECSA), supporting the SACC in its opposition to apartheid. A leading figure in ABRECSA as well as the UDF was Dr Allan Boesak, an academic theologian, prominent in the *NGK Sendingkerk* and member of the northern Cape family that was a pillar of the Carnarvon community. In the northern Cape, the Karoo Council of Churches was established as an affiliate of the SACC. Its offices were in Carnarvon, organised by Louise Boezak (Allan's aunt). Internationally, Allan Boesak played a leadership role in a meeting of the General Council of the World Alliance of Reform Churches in Ottawa in 1982, which declared apartheid to be a sin.[8] In the words of the World Alliance resolution:

> We declare with Black Reformed Christians of South Africa that apartheid ('separate development') is a sin, and that the moral and theological justification of it is a travesty of the Gospel and, in its persistent disobedience to the Word of God, a theological heresy.
>
> (de Gruchy, 2005, p. 193)

This resolution being labelled a '*status confessionis*' signalled the seriousness of the issue. The implication was that the credibility of the Gospels would be at stake if theological grounds were used to justify apartheid, and that to do so would be a heresy. Following this logic, if there was no theological justification, there was no religious basis for racial divisions to continue within the *NGK*. This was a position that many white church members found difficult to accept.[9]

At its 1982 meeting, the World Alliance of Reform Churches elected Boesak as its President. Following this, Boesak and others drafted the Confession of Belhar to give expression to the *status confessionis*. This was adopted by the moderating council of the Dutch Reformed Church in 1986, but given the highly sensitive nature of the Belhar Confession, the moderating council agreed that the decision on whether or not to include it as an article of faith would be devolved to individual churches. The church remained divided within itself.[10]

In Carnarvon, the whites-only *NGK* chose not to adopt the Belhar Confession and was determined to maintain its historically segregated status. The *Sendingkerk*, having joined the URC, pressed for change both within the church and more broadly. The URC congregations in Carnarvon and nearby Vanwyksvlei worked closely with the SACC in its opposition to apartheid. Reflecting on his experiences as a minister in the *VGK*, Sakkie Potgieter recalls this time as follows:

> The new initiatives of middle 1986 … inspired Louise [Boezak, aunt of Allan Boesak, and church member of the DRMC in Carnarvon] to once again take up the batten and become very active and contribute towards making the apartheid structures in Carnarvon unworkable. Pressure was mounting on all 'Coloured' representatives in apartheid institutions to withdraw their participation and thereby help rendering the town ungovernable. The NGO 'CARAVAN' [Carnarvon and Vanwyksvlei] was established to start addressing the socio-economic needs of the Carnarvon and Vanwyksvlei communities via alternative structures.
>
> In May, the executive committee of CARAVAN decided to invite dr. Allan Boesak, the world-renowned anti-apartheid activist and minister of the Uniting Reformed Church, and Rev. (at present Dr.) Lionel Louw of the African Methodist Church, to conduct church services in Carnarvon and Vanwyksvlei. Their arrival stirred up emotions amongst the local people….
>
> Enthusiastic, cheering youths clad in T-shirts of the United Democratic Front met the two invited pastors at the airstrip in Carnarvon and ran alongside the cavalcade all the way back to Carnarvon town, under the close watch of the South African Police and the security police. They were to conduct a morning church service in the neighbouring town of Vanwyksvlei first, before returning to Carnarvon for another service in the afternoon.

> In Vanwyksvlei, the church building of the URC was filled to the rafters and the reaction of the normally sedate and resigned people, was a revelation. Some community members, at a time when whites were still commonly addressed as 'baas' and 'nooi', were met by an explosion of built-up emotions. One man aggressively shouted to inquisitive whites: 'Ja, die waarheid maak seer, né?' (*Yes, the truth hurts, heh?*).
>
> For the afternoon service, a huge marquis tent had been erected on the open plain behind the URMC parsonage and opposite the Old Age home of Carnarvon. Some armed white commando members had taken up positions on the roof of the old age home, overseeing the proceedings, just in case they were called on to assist the police in restoring the order.
>
> The struggle for justice had now clearly reached new heights in Carnarvon. Carnarvon and Vanwyksvlei had taken up the challenge of establishing the Karoo Council of Churches, a regional affiliate member of the South African Council of Churches. I acted as its first chairperson, and Louise Boezak was appointed its Organising Secretary. Security police surveillance and intimidation started to intensify.
>
> There was a very strong awareness amongst those opposing the apartheid structures in town, that the crucial period in the history of Carnarvon town had come. The kairos, the moment of truth, the moment to take a firm stand, had arrived. This was indeed the period when the kairos document was first distributed throughout the country and on the streets of Carnarvon. Complicity by the church in these times was no longer possible. This was the kairos when peaceful civil disobedience and resistance had become the only remaining non-violent strategy left. The injustices of the past had reached a final boiling point and Carnarvon could not escape the heat. (Potgieter, n.d. SKA, pp. 67–68)

The Carnarvon *NGK* withdrew its promised stipend to Potgieter, who was then *VGK* minister in Vanwyksvlei, on the grounds that his sermons were too political, and the security police and army reservists in the region monitored church activities and harassed Potgieter's family, as well as prominent church members such as Louise Boezak and others in the community. During the 1986 state of emergency, the minister of the Carnarvon URC, Arné Leuvennink, was detained without trial in the local prison, alongside the principal of the primary school, Benjamin Hoorn, who was a minister in the African Methodist Episcopal Church (AME), and twelve young people.[11] As Hoorn recalls, 'ties between the two communities were broken'[12] – and they remain broken.

Sakkie Potgieter recollects some of his experiences as at the time as follows:

> My small wife (5'1" in the old language) who was threatened in the night in her own kitchen by big men in commando uniforms ...; security policy interrogation and intimidation; the burial of a thirteen year old

> boy who was shot dead by a '*kitskonsabel*;'[13] balled fists of the comrade pall-bearers; another arrest of a teenage girl who was dragged out from the burial assembly; the community pleading: *Dominee* [minister] go and talk to them. The community doesn't say it, but you hear it: *Dominee*, they are *your* people. And then those cold passages of the police station, the cold eyes and the icy answers.
>
> (Potgieter, 2017, p. 24, own translation)

In an interview for this study, the white woman minister of the *NGK* at the time raised the need for reconciliation between the two church communities. She repeated what other religious leaders in the towns had said: that reconciliation should start with the school. But why should the burden rest on the school? This question, and possible responses to it, will become clearer in the chapters that follow.

'The school'

Whether by explicit design or by default, the taken-for-granted assumption in the town in the early years of the twentieth century was that education would be provided in racially separate schools and on separate tracks. The taken-for-granted practice was that Coloured children would attend the state-aided mission school, primary only, on a non-compulsory basis. White children would attend public schools on a free and compulsory basis. Two tracks of schooling – mission and public – developed with different lines of provision, resourcing, and governance, and they afforded different opportunities to children according to wealth and racial designation (with gender not a major differential in attendance records in these years). Segregated schooling was established well before apartheid cemented the racial categorisation of schooling following the 1948 electoral victory of the Afrikaner-based National Party.

After the electoral victory of the National Party in 1948, apartheid legislation intensified the patterns of segregated schooling, establishing separate racially based departments. The notorious Bantu Education system brought the provision of schooling for black people under state control for the first time, and also established separate provision for people classified Coloured or Indian. Mission schools were closed or transferred to state control as required by the Bantu Education Act[14] and subsequently Coloured mission schools were moved under provinces and then separate Coloured Affairs control. Following the 1967 Education Act, schooling for whites was to be based on defined Christian and National principles, with different schools for English- and Afrikaans-speakers. Subsequent regulative arrangements made provision for education in Bantustans and for black people living in urban areas.

The legislation for apartheid education eventually spawned a host of racially and geographically based education departments: an overall national department of education; one national and four provincial departments for whites; one department for Coloureds; one for Indians; a national department for black people living in white-designated areas; and ten Bantustan departments – a total of 19 education departments.[15] These departments were funded at very different levels throughout the period of apartheid rule. By the time apartheid drew to a close, secondary schools had been provided in Bantustans and townships for people designated African, but schooling was not compulsory. This mix of arrangements formed the complex ground on which changes to education would need to be forged.

In Carnarvon, the mission school, classified Coloured, was initially transferred to the Cape province to administer before the 1963 Coloured Persons Education Act removed control of all Coloured schools to a separate department, cementing their separation from white education departments. In 1960, following a fire, the mission school was relocated across town and renamed *Laerskool* Carel van Zyl. In 1984, a secondary school was built in the Coloured township of Bonteheuwel across the dry riverbed from the town.

In 1987, a report by the Upington District Director to the Department of Internal Affairs (Coloured Affairs) motivated as follows for *Laerskool* Carel van Zyl to be moved to new premises on an earmarked site in Bonteheuwel near the secondary school:

> The present school building is in poor condition. The ceilings are busy collapsing and the classrooms appear untidy because the paint in most classrooms is peeling off. The building was last renovated and cleaned in 1979. Some of the windows are held together with wire.
>
> The lighting is very weak. Only nine classrooms have lighting, consisting of a single lightbulb. This makes the work of the students difficult, particularly in winter and on cloudy days.
>
> The office sections are inefficient and storage space is a great problem. The toilets are still a bucket system and must be emptied three times a week. This is very costly. The grounds are really uneven on one side and a lot of fill work needs to be done.
>
> In the light of the above it would benefit the Administration to construct the school and boarding house on the new premises. Costs will be saved because the site is very even, much closer to +- 80% of the school population and there are more opportunities for expansion.
>
> The school will be close to the Secondary School and therefore able to share power and water, itself a saving.
>
> It will also benefit the Administration with an eye to the future sewerage of the town. It will be very costly to lead a sewerage pipe from the present school, Carel van Zyl, to the sewerage-disposal dam, given that the land formation is stony. If the school is erected on the new premises only one pipe connection will be needed.
>
> The present site of Carel van Zyl does not lend itself to sports practice and the facilities in the town are also inadequate. The new premises provide many possibilities for expansion.[16]

Attaching this memo, the Chief Director of the Department of Local Government, Housing and Agriculture wrote a letter to the Executive Director of Education and Culture on 8 December 1987 about *Laerskool* Carel van Zyl:

> The school building consists of 40 classrooms and one office, with 13 classrooms being mobile units. The main building consists of 27 classrooms with plastered brick walls under zinc roofs and is in very bad condition, although structurally sound.
>
> Given the age of the building, the roofing, wooden floors, electrical installations, blackboards and notice boards need replacing.
>
> The estimated cost to upgrade the school to acceptable standards amounts to approximately R700 000,00. A new school for 1000 students and a boarding house for 200 students is planned for 1990, priority 189.
>
> Please indicate if you are in favour of an overall renovation, building of an administration block and toilet facilities. A definitive answer would be appreciated as soon as possible in order to make provision for funds in the new financial year 1988–1989.[17]

However, unexpectedly, records of the correspondence end at this point. Archival records for Coloured schools in the Cape close abruptly in November 1989 with the handwritten message '*Afgesluit op 30.11.89*' (Closed on 30.11.89) on a sheet of paper in the folder. It remains a puzzle that there is no further correspondence on the school, and no other archival records from relevant government departments. Staff at the Cape Archives have no knowledge of why the files were closed, or what happened to official records after this date. Given that this was a time when major political change was in the air, it is possible that these records for apartheid's Coloured Affairs departments were relocated, or warehoused, or destroyed, or simply lost. The reasons are obscure, but the results are clear: the interruption of a narrative mid-way, without a record of the next step, in a moment of change (Figure 5.1).

In fact, the next step in the narrative of the primary school was that it burnt down at the end of 1989. The fire was never investigated, and rumours circulate about possible causes: an electrical fault; naughty schoolchildren playing or smoking on the site; and even intentional arson, politically motivated. Some interviewees for this study were certain that they could identify the errant young men who might have destroyed the school without intending to; several white interviewees blamed the politicisation of youth at the time. Why was the fire not investigated? In a small and remote town at this time, the event was hardly even newsworthy. Without much infrastructural support in the town, perhaps an investigation into the loss of a Coloured institution at this point was not worth the costs to the municipality or the government.

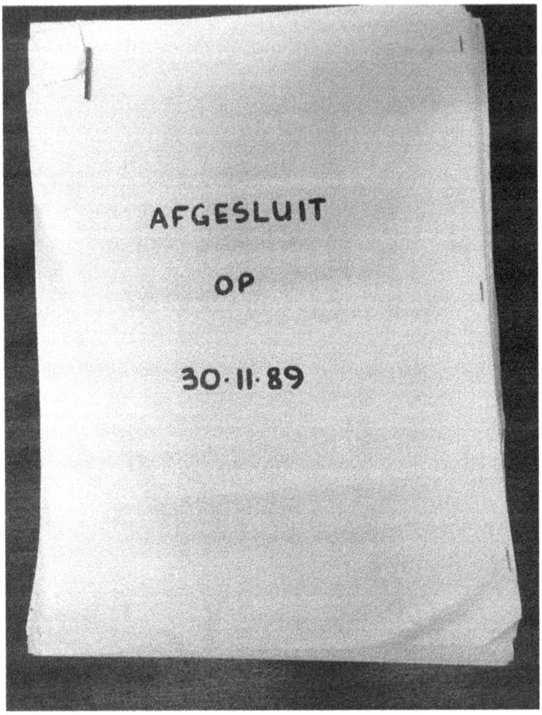

Figure 5.1 "File closed on 30.11.89".
Source: Western Cape Archives.

Whatever the cause of the fire, the consequences for the Coloured schools in the town were deeply unsatisfactory. A new primary school was not built on the earmarked site, and instead the staff and students were moved onto the premises of the secondary school in Bonteheuwel for the start of 1990 – where they remained as two schools sharing one set of premises until 2007. Why was a new school not built in 1990 on the designated site? Speculation is that there was a change of education districts and the paperwork was not carried through; or that in this period it was clear that school restructuring would be on the cards so nothing pre-emptive could be done; or that there was simply too much uncertainty about the future at that time.

Whatever the reasons, the result is clear: an erasure of social memory.

There is nothing remarkable about the Coloured schools in the small and remote town of Carnarvon that might suggest that any records were worth destroying, or that there was any particular motive in not investigating the fire at the school. Unimportance and neglect seem more likely reasons than any intentional action. Alternatively, it could be that Coloured schools in the Cape were engaged in boycotts and protests as part of larger social movements against apartheid at the time, and the records relating to Coloured schools were part of a larger archive that was of interest. Whatever the reasons for 'File Closed' in the Coloured Affairs archives highlighted by this case, the absence of files highlights a bigger issue of the destruction of public records by the apartheid government in the final years leading up to the political changes of 1994. Verne Harris (2000), an archivist in the Pretoria records management division of the State Archive Services from 1988 to 1994, has written about the widespread destruction of official records in the period leading up to 1994, and also compiled a report on this for South Africa's Truth and Reconciliation Commission in 1998. He makes the point that almost nothing has been written about archive destruction, and his account provides a chilling perspective on the fate of many records of social memory at this time. In an article on the destruction of public records by the South African state from 1990 to 1994, Harris makes the following comment:

> Under apartheid the terrain of social memory, as with all social space, was a site of struggle. In the crudest sense this was a struggle of remembering against forgetting, of oppositional memory fighting a life-and-death struggle against a systematic forgetting engineered by the state. The realities, of course, were a little more complex. Forgetting, for instance, was an important element in the struggles against apartheid – forgetting the half-truths, the distorted interpretations, the lies, of the apartheid regime. And the notions of 'oppositional memory' and 'state memory' themselves are problematic. They are artificial constructs, obscuring the sometimes fierce internal contestation in both spaces. Then there is the question of memory and imagination. Memory is never a faithful reflection of process, of 'reality'. It is shaped, reshaped, figured, configured, by the dance of imagination. So that beyond the dynamics of remembering and forgetting, a more profound characterisation of the struggle in social memory is one of narrative against narrative, story against story.
>
> (p. 29)

Much of the research on apartheid education has focused on the notorious Bantu Education Act of 1954 and its accompanying discourses of racism and inequality. Although the government's official line was 'separate but equal', the education acts that formalised segregation also embedded inequalities in every aspect of schooling provision. The rejection of 'Bantu education' (which came to stand for the whole system of racially divided and unequal schooling) was a focal point in the anti-apartheid struggle. Notable in the education struggle were the Soweto student uprisings sparked in 1976, the school boycotts sparked in the early 1980s in the western Cape, and the People's Education movement associated with the National Education Crisis Committee after 1986. The protests in education joined with the widespread struggle against apartheid, ultimately leading to its demise.

Concluding comments

From this brief outline of apartheid's architecture and the sketches of its enactment in the institutions of local government, church, and school from the vantage point of the small and marginal town of Carnarvon, it is possible to discern the structures of apartheid's racial capitalist state formation, and the brutality of its enforcement. Although Carnarvon's demographic and particular rural economy are not a prototype for other places in South Africa, they nonetheless provide a particular vantage point for magnifying a particular instance in the larger narrative of the spatial practices of apartheid's form.

The fact that apartheid was brought to an end by negotiated settlement rather than overthrow of the state meant the dismantling and replacement of an extensive legislative framework which crisscrossed all aspects of life – an extremely complicated process. This chapter provides mere gestures towards the complexity of the apartheid assemblage. Changing from apartheid involved not only its legislative network, but also shifting its hegemonic assumptions about difference, as well as structural arrangements in politics, economy, and society. These shifts – captured in the notion of the move from colonialism to decoloniality – are the subject of the following chapters.

Notes

1 In effect, the Bantustan policy was a way of administering difference as defined by ethnicity and crafted into spatial arrangements – presented as a form of self-government (and by extension, decolonization).
2 Interview, Louise Boesak, Carnarvon, 22 November 2018. The Boezak family applied for annual permits and were able to stay in their house until 1965. Potgieter (personal communication, 28 August 2019) notes that the following families lost their homes: Willemina Pietersen, Jan Daniels, Dampie Petoors, Jannie Witbooi, and Jan Links as well as of the Matthee, Christians, Snyders, Vass, Julius, and Andreas families. Given that this was based on memories of his congregants, he suggests that the list is incomplete.
3 Confirmed by his son, *ds* Andre Boesak. Potgieter, personal communication, 28 August 2019.
4 This point – together with the permit number – is recorded by Potgieter.
5 'Petty apartheid' refers to the multitude of racially discriminating practices of daily life, such as racially separate queues in banks and post offices, 'Europeans Only' benches in public spaces, separate public utilities, separate transport arrangements, etc.
6 There were other churches as well in Carnarvon, with a small, integrated Anglican church and the African Methodist Episcopal Church (AME) serving the Coloured community.
7 What makes this particularly ironic is that the *NGK* had been a founding member of the World Council of Churches in 1948. It broke its relationship in 1960 due to harsh criticism for its role in apartheid.
8 The *NGK* was a member of the World Alliance of Reform Churches, as well as the Reformed Ecumenical Synod, though by this stage it had left the World Council of Churches. (See de Gruchy, 2005.)
9 According to a number of interviewees, an implication of accepting the Confession of Belhar could be to impute that one's forebears were sinners.

10 Dr I. C. Potgieter, personal communication, 22 August 2019:

> The existing three confessions were *Die Nederlandse Geloofsbelydenis, Die Heidelbergse Kategismus, Dordtse Leerreels.* The only controversy arose from whether this document would become another theological declaration amongst many others, or whether it would be adopted as a fourth confession, which would be a very decisive step indeed.

11 Leuwennink was held for 42 days without trial, and Hoorn for 62 days. The DRC decided to cut its funding for Leuwennink's post and urged him to accept a transfer elsewhere. (Potgieter, n.d., SKA).

12 Benjamin Hoorn, Interview, Carnarvon, 26 January 2016.

13 A newly deputized constable in the South African Police Services during the last days of apartheid; they were noted for their zealousness and utter lack of police procedure. Their official name was 'special policeman'. https://wikidiff.com/constable/kitskonstabel 20190319.

14 The Catholics were a notable exception. They continued to run schooling in a dual system: mission schools for black students and fee-paying schools for whites, which subsequently began the 'open schools' movement by admitting black students after the 1976 Soweto uprisings.

15 These are sometimes counted as 15 national departments, plus 4 provincial.

16 Letter from Chief Director, Department of Local Government, Housing and Agriculture, Administration: House of Representatives, to Executive Director, Department of Education and Culture, 8 December 1987. TBK KUS-1615-7/11/10/ E 980-1.

17 Minute attached to letter from Chief Director, *op. cit., fn.* 16.

References

Cameron, R. (1993). Regional services councils in South Africa: Past, present, future. *Public Administration, 71*, 417–439.

de Gruchy, J. W., with de Gruchy, S. (2005). *The Church struggle in South Africa.* Minneapolis: Fortress Press.

de Visser, J. (2009). Developmental local government in South Africa: Institutional fault lines. *Commonwealth Journal of Local Governance, 2* (January), 7–25.

Harris, V. (2000). 'They should have destroyed more': The destruction of public records by the South African state in the final years of apartheid, 1990–94. *Transformation, 42*, 29–56.

Oomen, B. (2005). *Chiefs in South Africa: Law, power and culture in the post-apartheid era.* Oxford: Curry.

Picard, L. A. & Mogale, T. (2015). *The limits of democratic governance in South Africa.* Boulder, CO: Lynne Rienner.

Posel, D. (1997). *The making of apartheid, 1948–1961: Conflict and compromise.* Oxford: Clarendon Press.

Potgieter, I. C. (n.d., forthcoming). *The SKA in Carnarvon.* Unpublished mimeograph. (SKA).

Potgieter, I. C. (2017). Kan ek en Mariaantjie dáár saam skool toe gaan? In G. D. Cloete, A. Boer & J. S. van Rooy (Eds.), *Voor ons vergeet: sodat ons kinders ook kan weet.* Wellington: Christelike Lektuurfonds, 24–28.

South African Institute of Race Relations (SAIRR). (1950). *A survey of race relations: 1949–50.* Johannesburg: SAIRR.

6 Ending apartheid

In the crucible of the old, the new is formed

Negotiating change

In a spectacular moment in February 1990, Nelson Mandela walked free after twenty seven years in prison. Once deemed 'terrorist organisations', the African National Congress (ANC) and other political movements were unbanned by the apartheid National Party (NP) government so that negotiations for a new South Africa could begin in earnest. A mix of excitement and fear filled the air. A new order was possible, but it was not yet secured. What followed was a tense four years of negotiations between these erstwhile enemies to decide on a new constitutional order for the country – an outcome by no means certain at the time. Finally, in April 1994, South Africa held its first democratic elections, ending more than forty years of apartheid rule and more than three hundred years of colonialism. Two more years of negotiation were needed to finalise a new constitution in 2016.

The formal end of apartheid – heralded around the world as a victory for democracy and human rights – came at the same time as the crumbling of Europe's eastern bloc states, the fall of the Berlin Wall, and the fading of utopian visions of socialism that had inspired many in the liberation movements. In the ensuing context of neoliberal globalisation, South Africa faced the task of crafting its first democratic government. The arrangements made during this period were decisive in ending colonial logics of governance and laying the groundwork for alternatives.

In understanding the changes that were put in place, it is important to bear in mind that apartheid was ended by a negotiated compromise settlement between opposing forces, rather than by the 'overthrow of the state' in a decisive defeat. In the increasingly violent conflict between the apartheid state and the liberation movement, neither side could fully prevail – the apartheid state could not subdue its opponents, and the liberation movement could not overwhelm the apartheid state. An early agreement was that there would be a two-stage process of transition: a first stage where there would be a period of negotiation to draw up an interim constitution to allow for elections to take place; and a second stage where the newly elected parliament would act as a Constitutional Assembly to finalise the constitution.

This period was marked by tremendous uncertainty as well as continuing violence, and negotiations nearly stalled a number of times. Because of the extent of the popular struggles at the local level (which included local government, church organisations, and schools), parties agreed to hold a separate Local Government Negotiating Forum alongside constitutional negotiations. Early in the negotiating process, it became clear that 'traditional leaders' would insist on having a voice as well. And because of extensive protests and deep inequalities in education, new arrangements would need to be made for schooling – an issue where different views were strongly held during the negotiations.

The tough on-again-off-again negotiations of the all-party Convention for a Democratic South Africa (CODESA) and subsequent Multi-Party Negotiating Process (MPNP) dealt with many complex issues with regard to the structure and powers of new governmental arrangements. In the process of negotiations, the parties agreed on provisions for an interim constitution with a number of 'sunset clauses' to enable the first elections to go ahead. It was agreed that the first elected government would be a Government of National Unity (GNU), and that the positions of key apartheid bureaucrats would be protected for five years. The GNU of 1994 was made up of members of the ANC and its allies,[1] members of the apartheid National Party, and other political actors including the Inkatha Freedom Party and the Democratic Party. Negotiations lasted for another two years after the elections before the final constitution was adopted in 1996. Among the complex issues that parties struggled to reach agreement upon were property rights, labour lockout laws, and education (see Corder, 1994; Friedman, 1994; Segal & Cort, 2011; Strand, 2001).

The negotiated settlement established a constitutional democracy, in which citizens would have equal rights before the law for the first time in South Africa's history. Embracing a constitutional democracy meant that the repeal of apartheid laws would take place through parliamentary procedures, and the rule of law would prevail. As part of the negotiations, nine new provinces were demarcated from the previous four provinces (with Carnarvon falling in the newly constituted Northern Cape Province), and a whole new sphere of government was introduced at the local level (see Map 6.1). Existing apartheid legislation was amended; new laws were passed; and government departments were restructured to include a mix of old bureaucrats (who knew how the system worked to enable change, or to block it) and new appointees (many of whom had struggle credentials but little or no experience in government or administration). It was in this context that a raft of new education policies was assembled, with complex and contradictory consequences, as described in the following two chapters.

However, while constitutional arrangements guaranteed democracy and equal rights for all, unresolved issues have remained contentious. Notably, little was done to shift patterns of economic ownership, though the ANC-led Alliance had alluded to this in its early days in government. Social grants and subsidies for basic services have ameliorative effects, but do not provide

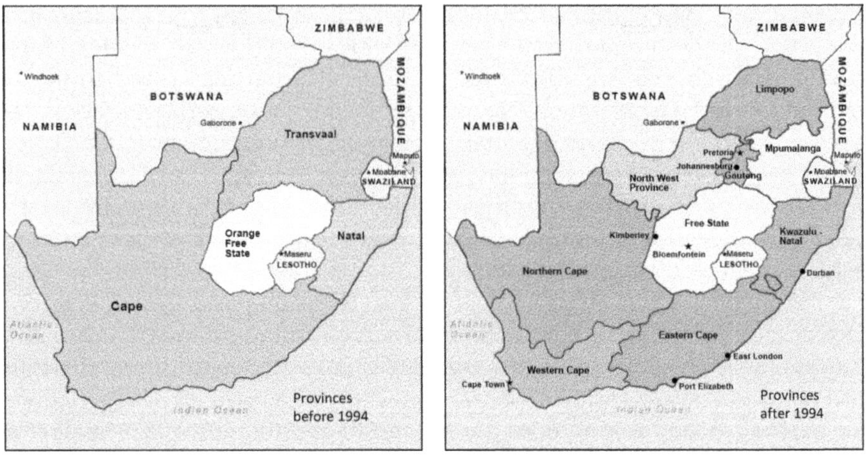

Map 6.1 Provinces of South Africa before and after 1994.

a means for wealth to shift. The constitution recognised property rights but left the contentious issue of land redistribution as an open issue which remains to be addressed. Another unresolved issue was the role of traditional leaders, whose power was anchored in Bantustans and who staked their importance on their capacity to marshal rural votes.[2] A separate chapter of the final constitution gave formal recognition to 'the institution, status and role of traditional leadership according to customary law … subject to the constitution'. This ambiguous wording indicates that the structures of indirect rule under colonialism and apartheid still held valence. Over time, legislation has expanded the powers of traditional authorities, restricting the land ownership and citizenship rights of rural people living in former Bantustans. As Aninka Claassens (2014) notes:

> traditional leadership laws … entrench the 'tribal' boundaries which make up the former homelands, and recent policies … foreclose land-ownership for the majority of rural people. I argue that these laws and policies reinforce, rather than address, the legacy of the 1913 and 1936 Land Acts. Distorted constructs of unilateral chiefly power are mobilised in attempts to create a separate legal zone of customary authority that undermines the citizenship rights of those living within the boundaries of the former bantustans.
>
> (p. 761)

As mentioned earlier, the town of Carnarvon was historically not part of any Bantustan; if it had been, this book would have given more prominence to this issue, since it is central to an unfolding decolonial politics.

In short, although the political changes of 1994 were seismic, they also left certain contentious issues unresolved, and the accompanying economic changes were not of the same order of magnitude. Whereas the Reconstruction and Development Programme (RDP) introduced in 1994 was intended to achieve 'growth through redistribution', this more radical approach was dropped in 1996 in favour of the Growth, Employment and Redistribution (GEAR) strategy. GEAR was an overtly neoliberal economic strategy intended to align South African markets with the global economy, and its thrust was distinctly incompatible with the equity strategies of the RDP. In effect, these economic arrangements ratified the existing distribution of wealth and left the structure of ownership much where it was under apartheid. The introduction of measures for Black Economic Empowerment (BEE) and pro-poor social grants softened but did not shift the existing patterns of wealth and ownership of resources. The overall result was limited socio-economic restructuring during and after the transition (see Bond, 2014; Marais, 1998; Seekings & Nattrass, 2015; Terreblanche, 2011).

As part of the transition process, the Truth and Reconciliation Commission (TRC) was set up to acknowledge some of apartheid's worst atrocities. Important though this was, it was a limited process that could not address the extent of the profound damage and human suffering that apartheid had caused. The TRC conferred sufficient legitimacy for the new government to break with the apartheid past – without fully addressing its devastating effects on ordinary people's lives.

For optimists, the suite of changes that was achieved through negotiation represented the launch of the 'Rainbow Nation', but perhaps it would be more accurate to say that the project of national unity was nowhere near completion. It was a project still to be built – a point made particularly well at the time by Neville Alexander's (2002) essays in *An Ordinary Country*. More than two decades after the elections, with apartheid's legal framework largely removed and the institutions of political democracy in place, there are still deep social and economic inequalities that divide the nation. The templates of apartheid and earlier colonialism are still evident in the social structures of the economy as well as in multiple practices of daily life.

Post-apartheid South Africa is marked by continuing problems of high poverty, high inequality, and high unemployment.[3] Official statistics show that half of the adult population live below the upper-bound poverty line, with adult women experiencing higher levels of poverty than men (StatsSA, 2019). In their study of poverty and mobility over ten years, Zizzamia, Schotte, & Leibbrandt (2019) have shown that households that are chronically poor (meaning that they are unable to meet basic needs) have little chance of moving out of poverty (p. 19).[4] They are typically concentrated in rural areas where they are structurally excluded from labour market opportunities; they are predominantly African; and most have not completed secondary education. Female-headed households are more likely to slip into poverty and less likely to escape poverty. Importantly, the authors note that 'While social grants are key to the survival of the chronic poor, they do not address the structural barriers to upward mobility' (p. 25). What would be needed, they argue, are opportunities to break structural barriers, and strategies for this require more extensive research on poverty dynamics.

Descriptions of inequality in South Africa are well set out in two major reports: the World Bank's *Report on Poverty and Inequality in South Africa* (2018) and the *World Inequality Report* (Alvaredo et al., 2018) by a group of economists including the well-known Thomas Picketty. Both reports acknowledge that social policies have ameliorated, but not transformed, structural inequalities in South Africa. The *World Inequality Report* (Alvaredo et al., 2018, p. 149) notes that South Africa is one of the most unequal countries in the world but also that inequality has increased significantly since 1994 in spite of pro-poor interventions and that top-income groups are still overwhelmingly white. The World Bank's (2018) *Report on Poverty and Inequality in South Africa* states that:

> The 'social wage' has been used as a redistributive mechanism of the government budget deliberately aimed at improving the lives of the poor and reducing their cost of living. This has been achieved through, among others, free primary health care; no-fee paying schools; old age and child support grants; housing; and free basic services (water, electricity and sanitation) to poor households. Although these policies and interventions have resulted in notable gains in poverty reduction since 1994, the country continues to face the challenge of high poverty, high inequality and high unemployment. The persistence of these challenges calls for a rigorous assessment of the drivers, constraints and opportunities for poverty and inequality reduction in South Africa.
>
> (2018, p. x)

The spatial dispersal of poverty, and the degree to which it has not shifted, is outlined as follows by the World Bank Report:

> Poverty has a clear spatial dimension and spatial patterns of poverty suggest progress toward dismantling the spatial legacy of apartheid has been slow. Rural areas remain the regions of highest poverty concentration. The results reveal a notable divide in poverty levels between two sets of provinces: Free State, Gauteng, and Western Cape versus Eastern Cape, KwaZulu-Natal, and Limpopo. This divide is a clear legacy of apartheid: compared to Eastern Cape, KwaZulu-Natal, and Limpopo; the Free State, Gauteng, and Western Cape did not have high concentration of "homelands" during apartheid. Homelands were areas set aside for black South Africans along ethnic lines during apartheid. Public service delivery and infrastructure was poor in these areas.
>
> (2018, p. 41)

As well as poverty and inequality, South Africa has high unemployment rates: at the time of the 2019 elections unemployment was officially 26.7% and unofficially much higher. Youth unemployment in 2019 was particularly high: 55.2% for ages 15–24 and 63.4% for ages 15–34. The labour market continues to discriminate on lines of race and gender, and more women are in long-term unemployment than men (StatsSA, 2018; see also Bhorat & Goqa, 2013). Quality of schooling also influences labour market opportunities, as does level of education attained. Moreover, because of shifts within the labour market itself, success depends on having at least completed matric – a criterion that many black people do not meet (Branson, Garlick, Lam, & Leibbrandt, 2012).[5]

The tensions between deeply entrenched socioeconomic inequalities and the symbols and institutions of a modernist democracy have brought continuing social instability in South Africa, involving both new black elites and protest movements at the local level. For new elites, there are limits to economic empowerment, while for poor people, social grants from the state do not satisfy aspirations. The tensions are particularly clear in the case of local government, which also illustrates the enormous complexity of changing the government while running it. Although the focus of this study is on schooling and place, governance arrangements at local and provincial levels (including the links between local institutions such as municipality, school, and church) provide an essential context for understanding how the changes of 1994 took shape in Carnarvon in the years that followed, and will be covered briefly in the following section.

Difficulties in establishing local government

As discussed in Chapter 5, local government was a flashpoint in protests against apartheid, and its confused mix of structures and distorted funding arrangements meant that major structural changes were required. Consequently, a Local Government Negotiating Forum was been established alongside broader constitutional negotiations during the transition period, and the Local Government Transition Act (1993) set out a phased approach to local government changes. Plans were made for transitional local councils to be elected in 1995/6, and full-fledged elections were scheduled for 2000. Interestingly, even though over 40% of the population live in rural areas in former Bantustans, traditional leadership was not included in these local government arrangements, with the powers of traditional authorities recognised separately in the constitution.

New constitutional arrangements for governance had important effects on towns like Carnarvon. Changes to provincial boundaries and the introduction of a separate new sphere of local government brought policy shifts that rippled into local dynamics and destabilised them. Once again, differences are striking when the narrative of change is told from the peripheral vantage point of a small and marginal town, rather than following master narratives from the centre of state power. The following sections look at the complex restructuring of local government and the resultant difficulties, and also how local government changes played out in Carnarvon.

Local government changes

The opening words of the new White Paper on Local Government (Department of Provincial and Constitutional Affairs, 1998) leave no doubt about the 'historical role of local government in creating and perpetuating local separation and inequity', the significance of local resistance in political struggle, and the importance of change at this level:

> Apartheid has fundamentally damaged the spatial, social and economic environments in which people live, work, raise families, and seek to fulfil their aspirations. Local government has a critical role to play in rebuilding local communities and environments, as the basis for a democratic, integrated, prosperous and truly non-racial society. (Preamble)

The new constitution ambitiously envisaged a complete transformation of local government that would enable local communities to participate in democratic decision-making, and at the same time build a local developmental focus. New legislation made provision for a 'wall-to-wall' system of elected municipalities with responsibilities for integrated development planning as well as administration and service delivery.[6] By 1999, interim arrangements saw some 12,000 local authorities reduced to 843 transitional municipalities, and these were further reduced to 284 for the elections of 2000.

The Constitution set out a transformative vision, where a standardised set of institutions across the country would take responsibility for the local government with an extensive mandate: service delivery; setting up democratically elected structures and harnessing the participation of local communities; and meeting the developmental mandate set out by the Constitution. This range of tasks has proven almost impossible to implement, with constrained funding being a major problem from the start. The one-size-fits-all approach has been particularly problematic given the extreme inequalities between municipalities and the limited resource-base of many (if not most) of them, with difficulties made more acute by the neoliberal approach of market delivery, fiscal restraint, and performance management.

There is an extensive and growing literature appraising post-apartheid local government changes: the administrative structures and operational requirements; the demands of the new approach to planning; and the development challenges facing the Karoo region and its small towns (see Atkinson, 2007; Picard & Mogale, 2014; Powell, 2012; Rogerson, 2011; SALGA, 2015; Siddle & Koelble, 2012).[7] Much of this literature highlights the near impossibility of achieving success in this new sphere of government.

Perhaps the most sobering assessments are to be found in the annual reports of the Auditor General (2017) on the operations of district and local municipalities, showing that only 13% of municipalities received unqualified audits in 2016–2017. The 2016–2017 Report outlines likely reasons for failures of accountability, including: vacancies and instabilities in key positions; inadequate skills, particularly in terms of oversight of financial and performance management; political infighting; an attitude of 'blatant disregard' for controls and compliance with key legislation in some municipalities; leadership not taking recommendations and warnings seriously; and inadequate support for municipalities from provincial and national role players. That the national government itself views local government performance in such a poor light is a striking indication of its shortcomings.

Mismatch between political and economic changes?

It is ironic that demands for greater participation in decision-making in popular struggles against apartheid have resulted in cumbersome administrative structures and over-ambitious development planning tasks, with changes too elaborate to be implemented as envisaged. The results have been complex, including multiple increasingly violent protests at the local level; the emergence of new social movements; and elite involvement in local maladministration and corruption.[8]

Peter Alexander (2010) describes local government protests as 'a rebellion of the poor', summing up his position as follows:

> The protests reflect disappointment with the fruits of democracy. While some people have gained, the majority are still poor. Levels of unemployment are greater than in 1994, and income inequality remains vast. People can vote, but all too often elected representatives are self-seeking and real improvements are few. Many problems can be traced back to post-apartheid government policies that can be described as 'neoliberal'. Privatisation of local services opened up new opportunities for private accumulation by councillors and their cronies.
>
> (p. 37)

Overviewing social movements in post-apartheid South Africa, Richard Ballard (2005) writes:

> The new generation of social movements appeared in earnest once the ANC's second term in office began. The Treatment Action Campaign (formed in 1998), Anti-Eviction Campaign, Anti Privatisation Forum, Soweto Electricity Crisis Committee (2000) the Landless People's Movement, Coalition of South Africans for the Basic Income Grant (2001) and the Education Rights Project (2002), have been amongst the

> more enduring and visible struggles to have reconstituted a vibrant oppositional civil society. Countless unnamed small scale and ephemeral struggles have also emerged across the country.
>
> (p. 83)

Local government struggles have a significant gender dimension, though as Allison Goebel (2011) rightly points out this has tended to be a 'blind spot' in analyses.[9] Goebel notes that most participants in grassroots protests are women, even though movements are mostly led by men, and she makes a convincing case for examining the gendered ways that the poor are linked to the state through grants and other services, as well as the gendered dynamics of popular protest.

The involvement of new black elites jockeying for access to political and economic influence has also been a destabilising occurrence at local level. This is well set out by Karl von Holdt (2013), who argues that these elites have forged shifting links with local groups organised around concrete grievances, with each group using the other to further their interests. Von Holdt argues that new elites, faced with limited opportunities for black economic empowerment in the private sector, have turned to state institutions to provide opportunities for income streams. These elites contest between themselves for positions of power and influence, for state contracts, for the award of tenders, and for access to the distribution of employment opportunities. In this fierce competition, there is widespread 'collusion between state officials and contenders for contracts in order to rig the tender process and inflate contracts'. Because this entails breaking the law, 'the result is intense struggle for the control of the state institutions responsible for the rule of law' (p. 594).

Given the importance of von Holdt's (2013) argument and its complex implications, it is worth quoting a section in full:

> The subordination of state, legal and constitutional institutions to the goals of personal and factional enrichment and accumulation is frequently condemned in the public arena as 'corruption', or the criminal activity of a small number of rotten apples. However, these practices are so pervasive that they amount to collective and tacitly acceptable practices in wide political circles ….
>
> There are continuities and discontinuities with practices in the apartheid state. Under apartheid the state was a site of power that privileged white and specifically ethnic Afrikaner employment (Posel, 1999); and that facilitated the formation of white Afrikaner capital. There was also patronage and corruption, but these were generally successfully concealed from the media. The rents generated by Afrikaner control of the state and ethnic mobilisation were directed to productive investment and capital accumulation. All of these practices were consistent with the apartheid symbolic order, serving to deepen and strengthen it.

> In contrast, intra-elite conflict and the elite moral order of personal enrichment in post-apartheid South Africa tend towards the dislocation of the democratic symbolic order and are in many cases directed towards conspicuous consumption rather than productive investment. Intra-elite conflict ramifies through party, state and society, is unstable, publicly visible and increasingly violent, in contrast to the practices of domination and control within the apartheid regime.
>
> (p. 594)

A different form of organised protest at the local government level – and one that has occurred in Carnarvon – is the withholding of rates by local (white) ratepayers to signal their dissatisfaction with new administrations. Powell (2012) describes these ratepayer protests as follows:

> In towns across the country, ratepayers associations have declared disputes with municipalities over poor service delivery, corruption and mismanagement, withheld the payment of rates and taxes to their municipalities in response and, in some cases, assumed the responsibility to provide municipal services.
>
> (p. 23)

Local government protests continued, an expression of justifiable anger at the broken promises of change. The difficulties of bringing about change at the local level highlights the messiness of apartheid's legacy, the over-ambitious scope of changes envisaged by the new government, and the tension between political and economic arrangements in the post-apartheid settlement. Alarmingly, the corruption spawned by this contradictory settlement has severely compromised the moral order of the new state as von Holdt's (2013) work shows. Taken together, these conditions exemplify conditions of coloniality, where complex intersectional inequalities remain after colonial administrations are dismantled.

The next section looks at how these new governmental arrangements played out in the town of Carnarvon, with its historical patterns of possession and dispossession, and its particular social institutions.

Carnarvon in the era of development planning

New mappings

The new provincial boundaries and local government demarcations meant that the spatial relations of Carnarvon were mapped in very different ways. The restructuring of provincial boundaries placed Carnarvon in the new Northern Cape Province, looking north to its capital city of Kimberley, instead of south to Cape Town as it had done in the past. The Northern Cape Province covers the largest area in South Africa and has the smallest population. In the new local government arrangements, Carnarvon changed from having its own dedicated whites-only municipality, to being grouped with two other towns (Vosburg and Vanwyksvlei) in the Kareeberg Local Municipality, one of eight municipalities in the Pixley ka Seme District Municipality, which in turn was one of five district municipalities in the Northern Cape Province (see Map 6.2). With the seat of the District Municipality in De Aar, Carnarvon found itself on the western edge of a large local government area. The nearby towns of Victoria West and Williston – both historically significant reference points for Carnarvon – each fell under different district municipalities.

Under the new mapping, the geographical area of the Karoo was divided between four different provinces (Northern Cape, Free State, Western Cape, and Northwest Province). This remapping reflected political rather than environmental logic, given the arid ecology and fragile economic affordances of the Karoo. In each of these provinces, the Karoo was mapped as a marginal zone in development terms.

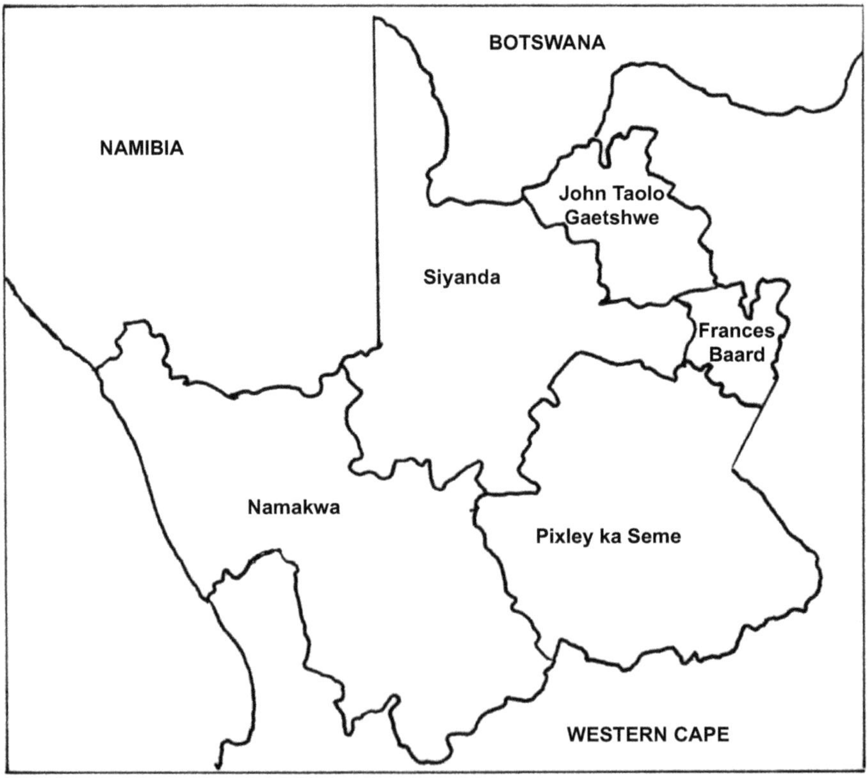

Map 6.2 Northern Cape district municipalities.

The Northern Cape Province has the largest geographic spread of the provinces, and the smallest population, with towns situated 80–100 kilometres apart (Atkinson, 2019). Its arid environment supports low intensity farming (mainly sheep and goats) and more recently, game farms. Nonetheless, its towns have relatively sophisticated infrastructure in terms of road, telecommunications, electricity, and so on, and there are good communication links with South Africa's major urban centres. In terms of population profile, 40% of the population is classified Coloured by the Census of 2011.

Carnarvon's 'municipality'

Prior to 1994, the town of Carnarvon had its dedicated municipality and Coloured management committee, with surrounding farms falling under Cape provincial arrangements, and its resources were channelled through a Regional Services Council. Many white farmers had connections to the town, including a town house and membership of the church, and they also voiced their collective interests through Farmers Associations or Agricultural Unions. De Visser (2009) provides a clear description of the changes to these arrangements:

> Before 2000, the average municipality was governed by a small council with a weak, collective executive structure. The council was chaired by a mayor whose task was largely ceremonial. The municipal administration was led by a strong 'town clerk' who initiated and drove much of the council agenda (Olowu & Wunsch, 2004, p. 89). This fitted the context of the municipality as a largely administrative, rather than policy making, authority.
>
> The new generation of municipalities is governed by a large council; it has a strong executive authority, in many cases concentrated in an executive mayor. The council meeting is chaired by a separately elected speaker ... and the administration is headed by a municipal manager. The municipal manager is appointed by the council and is expected to work very closely together with the municipal executive. The new legal framework expects municipalities to extend their activities beyond administering national and provincial laws. Municipalities must adopt policies and by-laws, actively engage municipal communities, plan strategically and partner with external institutions. This requires strong political and administrative leadership. A critical difference from the system that prevailed before 2000 relates to the role of the municipal executive. The municipal executive is expected to initiate policy, oversee the administration and take regular executive and administrative decisions.
>
> (pp. 14–15)

It is these changed arrangements that white townspeople referred to in their complaints about having 'their' municipality taken from them. Gone were small town municipalities serving white communities exclusively, and in their place were large district and local municipalities with elected councils and salaried administrative officers in a completely new, separate sphere of government. Moreover, according to a former mayor, the three towns of Carnarvon, Vosburg, and Vanwyksvlei were none too pleased about being put together under a single municipality of Kareeberg 'without having a choice' and they grumbled between themselves.

While '*the municipality*' in its apartheid form might have been romanticised by sections of the white community in Carnarvon, this was hardly a shared sentiment. The majority of community members had no say in the municipality under apartheid, and no reason for fond attachment to past arrangements. In fact, anti-apartheid resistance featured prominently in the town's history, as mentioned in the previous chapter. Carnarvon had its own struggles against apartheid – not as violent as other places in the protest movements of the 1980s, but certainly marked with its own stamp. Interviewees told a number of stories of the 1980s: of political meetings in the town, including United Democratic Front (UDF) protest meetings against apartheid; of the symbolic importance of Allan Boesak, a pastor in the *Sendingkerk*, leader of the UDF and member of the extended Boesak family of the northern Cape; of the presence of local reservists of security police threatening potential anti-government protestors; of attempts to intimidate the pastor and family of the *Sendingkerk*; of the arrest and detention of the *NGK* pastor and primary school principal. Louise Boesak from the town had been elected to the Coloured Representative Council in 1980 in the 'own affairs' constitutional arrangements prior to 1983 – with the express purpose of closing down separate representative bodies for Coloureds in a protest vote (a move which was not achieved, with the Tricameral Parliament being set up in 1984). In 1994, Louise Boesak was elected by the ANC as member of the first parliament of the Northern Cape Province. These examples of activism in the town illustrate the local forms in which broader political struggles were played out.

Over the years, election results showed shifting political alliances. In the first local government elections of the Kareeberg local government, the newly constituted Democratic Alliance (incorporating the former National Party) won the majority of the seats, but in 2006 they were supplanted by the ANC.[10] Interestingly, in 2016 the Economic Freedom Front (EFF) first made its presence felt in the ward of Carnarvon (Tables 6.1 and 6.2).

A brief quote from one of the townspeople interviewed in this study provides a particular perspective about the presence of the opposition Democratic Alliance party in the countryside:

> What is worrying about the DA in the countryside – not necessarily in the cities – is that we know who the people are who were absolutely opposed to anyone in the struggle. They were the opponents and they never apologised for anything or gave any idea that they had changed. Now they are all DA. Why? Because they are the only party to oppose

Table 6.1 Kareeberg local municipality election results

	ANC%	*DA%*	*EFF%*	*COPE%*	*IFP%*
2000	49	50			3
2006	63	31			3
2011	61	32		7	
2016	46	31	12	2	

Source: Own calculations from Electoral Commission website.

Table 6.2 Election results for the Carnarvon ward

	ANC%	*DA%*	*EFF%*	*COPE%*	*IFP*
2000	45	55			
2006	55	36			
2011	51	35		12	
2016	34	37	13	5	

Source: Own calculations from Electoral Commission website.

> the ANC. They're everywhere here in Carnarvon, and that's a problem for the party. I don't know what they can do about it.

From the perspective of some of the interviewees in this study, the new and expanded system of local government was a point of resentment. Local government seemed bloated, particularly compared with the former white 'niche municipality' of apartheid days. In this vein, de Wet Nel (2015) writes:

> The advent of democracy in 1994 resulted in the municipality becoming a political bureaucracy and almost immediately a source of income for a number of people from one political party.
>
> Councillors who were never paid before now, were suddenly given the power to determine their own salaries with the result that exorbitant salaries disproportionate to their qualifications, capabilities, experience or expertise were introduced for councillors and staff where politics was the main consideration in appointments.
>
> (p. 85)

It is certainly the case that many local white people acutely expressed their loss of voice at the municipal level. Once a support, the government and municipal structures had become a hindrance to their interests. After the local government town moved from DA to ANC majority, the (white) Carnarvon Rate Payers Association was formed in 2004 to give voice to concerns about the town's management, and between 2006 and 2011 a group of rate-payers withheld payment over allegations of mismanagement, putting money into a trust account instead.[11] The group lost a court bid against the municipality. Several white interviewees complained that members of the new structures 'had no clue of how the system worked', that 'proper policies were not followed', that 'people who haven't managed their own accounts are then given control over the town's accounts', and 'people don't even know when they're doing wrong'. None of these interviewees linked this lack of knowledge of procedures with the racial exclusivity of the apartheid governance structures they had supported. Political division and suspicion are pervasive in the town, whether as complaints about politicisation, service delivery, alleged corruption, or general mismanagement – points affirmed by the research of the Centre for Development Research (2010). Clearly, changing administrative structures is difficult in itself, but shifting attitudes of entitlement and superiority is another matter.

Profile of the town

The Census of 2011 provides a particular representation of the town of Carnarvon, sketching its political economy and social conditions as follows:[12]

- Carnarvon had 6,612 residents, with the total population of the municipal district of Kareeberg being 11,673.
- There were 3,222 households in the Kareeberg, of which 685 were agricultural households.
- Nearly 90% of Carnarvon's residents lived in formal dwellings.
- Of the total population of the town, 26% were not economically active, and of the economically active population, 25% were unemployed. Youth unemployment stood at 32%.
- Nearly half of the population (43%) lived in poverty.
- In the over-20 age group, 18% had no schooling, and a mere 17,5% had matriculated. There was no provision for post-schooling education and training.
- More than three quarters of residents (77.4%) had no Internet access from home, work or elsewhere.
- In population classification terms, 85.1% were Coloured, and 9.1% white. The vast majority (93.7%) were Afrikaans-speaking.
- There was a slight increase in Carnarvon's population between 2007 and 2011 but no marked pattern of in-migration of black people (unlike some of the other small towns in the Northern Cape).

Added to this, over one-third of the population in Carnarvon received a grant or family support of some form in 2008 – a significant proportion of people.[13] Research by the Centre for Development Support (2010) suggests that grants play a complex role in communities such as this: on the one hand, 'grants seemed to have been playing a substantial role in lifting people out of destitute poverty', and 'probably played a crucial role in reducing urbanisation and migration away from these towns' (p. 51). On the other hand, 'Grants, despite their positive contribution in reducing poverty, unfortunately also held the negative implication of increased state dependency' (p. 51).

Development plans and marginalisation

The approach to development planning adopted by the post-apartheid government has served to underline the marginalised position of small Karoo towns such as Carnarvon. Development planning is intended to be a function of all spheres of government, and the National Spatial Development Perspective (NSDP) developed by the Presidency provides the national masterplan for this approach (see Nel & Rogerson, 2009). It sets out categories of potential development across the country and provides guidelines and appropriate interventions for different economic regions. Areas with identifiable economic potential for development and job creation are prioritised; activity corridors and nodes linking to areas with high potential for growth are targeted for investment; and marginal areas are to be provided welfare support and limited interventions.

That approach to development confirms the marginalised position of the Karoo is argued by Nel and Hill (2008) as follows:

> In contrast to the government focus on farming concerns in the 20th century, in recent years the post-apartheid government is adopting a policy with a strong urban focus. This reflects the reality that the majority of the population has now urbanised and that the economy is dominated by the larger cities. In policy terms, the key 2003 National Spatial Development Programme divides the country into areas of high and low resource potential and economic activity and they contain proposals that government resources and support focus on the core nodes (The Presidency, 2003). Under such a scenario, the extensive rangeland of the Karoo with its low population density has been systemically marginalised.
>
> (p. 2669)

In terms of spatial development planning, all provinces, districts, and local municipalities are required to draw up integrated development plans (IDPs) following a common framework, in order to inform specific planning, budgeting, and decision-making. This is a large task, and a brief reading of the different plans suggests that it is more straightforward to shift the discourse towards development planning than it is to bring new insights or strategies to light. The Northern Cape Province Spatial Development Framework (NCPSDF) classifies Carnarvon as a 'service centre' that

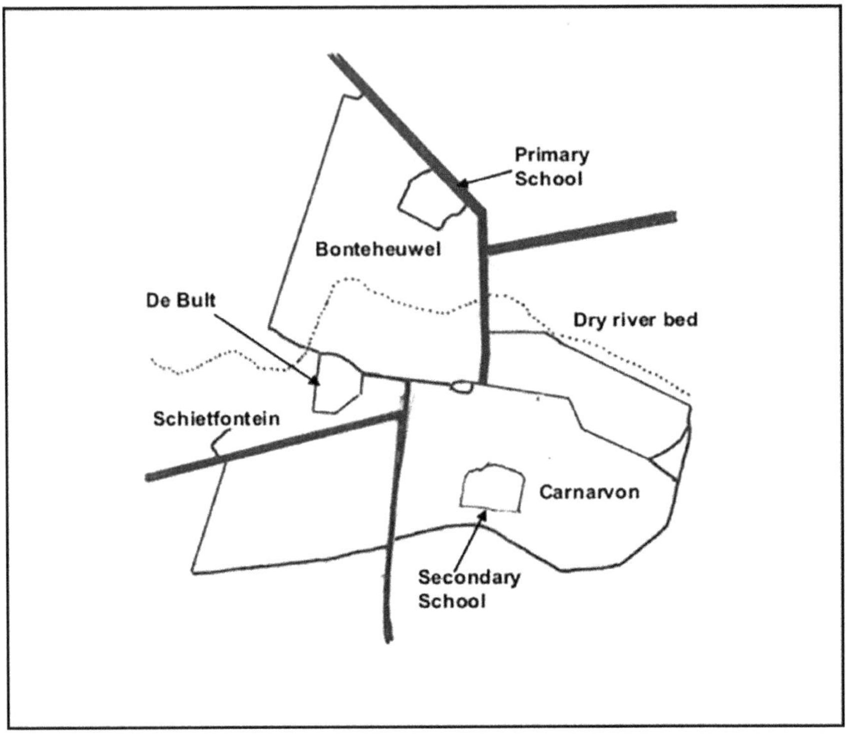

Figure 6.1 Layout of the town of Carnarvon.

is 'transitional' in the quadrant structure of high/low need, and high/low development potential. In other words, it could be said that this is not regarded as a place that is expected to yield much either way.

The overlapping (if not repetitive) local economic development plans of Pixley ka Seme District Municipality and the Kareeberg Local Municipality appear to be written to template-type requirements, with little mention of actual planning initiatives in specific places. What their form and content indicate is just how difficult it is to bring about developmental changes in a marginal region where resources and expertise are in short supply.

The economy of Carnarvon

The Northern Cape Province Spatial Development Plan (2012) notes the following:

> In many towns segregation and disparities inherited from the apartheid era are still visible in various forms, especially in spatial and institutional structures, as well as in some residents' perceptions and

> behaviour. Intentions and resolutions in this regard are generally well documented, but implementation is slow. This is probably one of the reasons for the persistence of high levels of human needs in many towns. (p. 106)

In Carnarvon there is little evidence of development. Divisions of race and class are all too visible in the streets and houses of the town itself. In the town, middle-class areas have been racially porous to some degree, but the poverty of the historically Coloured areas of De Bult, Bonteheuwel, and Schietfontein is striking as well as their continuing segregation. The continuing existence of two racially separate Reformed churches around the corner from each other signifies that the historical divisions and spatial practices of race are still enacted in the religious life of the town. Ownership of the town's private resources such as farms and businesses remains in white hands. Local government – now a source of employment including well-paying jobs for a few individuals – is the one area that has visibly shifted away from all-white control. Local council elections have been hotly contested over the years, and local government has passed from DA to ANC hands.

Overall, the patterns of resource distribution in the town provide a salutary reminder of the benefits and limits of political change. The concentration of ownership of economic resources in white hands and the high levels of unemployment and poverty in the Coloured community show continuities with the past, ameliorated but not removed by the policy and development approaches of the post-apartheid government.

The infrastructure of small Karoo towns, such as Carnarvon, reflects their historical establishment. Atkinson (2007) notes that the infrastructure in these towns is generally good, representing 'a great deal of sunk capital in housing, water, sanitation, roads and other infrastructure' (p. 8). So, in Carnarvon, most people occupy formal houses, with running water and electricity available and subsidised for the poor, and most people have access to schools, social service offices, the clinic, and shops. Although 43.5% of people are classified as living in poverty, this poverty has a different 'face' to the poverty of urban shack lands or rural former Bantustans.

That said, there are few economic opportunities outside of agriculture and services and as with other small towns, there is a tendency to lose purchasing power to larger regional towns. Indeed, a senior official in Carnarvon interviewed for this study suggested that one of the consequences of white children attending schools in other towns has been the depletion of middle-class recreational and even church activities in the town, as parents travel to other towns to visit their children, attend church, and purchase their supplies.

The main economic activity in the area is sheep and goat farming – wool as well as meat marketed as 'Karoo Lamb'. Farming practices have changed over the years, with incremental but powerful effects on employment and residence patterns. Changes include the amalgamation of farms to form larger units, limits on stock to prevent overgrazing, and fencing (while not always maintained) for protection of animals against black-backed jackals and small feline predators. The natural resources of the place (particularly water) set limits on farming activities, and the shift from wool to meat farming has meant a reduction of jobs on Karoo farms and the movement of people to towns. Owners do not always live on their farms, since they may have several farms as well as a house in the town. Sheep farming in Carnarvon is not labour intensive, we are told, and does not offer extensive employment opportunities. As one of the local townspeople remarked, 'God looks after the rain and the sheep look after themselves'. Mention was also made of the Extension of Tenure Act (ESTA), with farmers possibly considering it easier to buy a house in the town for a valued worker rather than risk their having a legal right to live on the farm.

The Centre for Development Support (2010) estimates that between 30% and 40% of jobs in the primary sector were lost in Carnarvon and its surrounding towns between 1996 and 2005 – a marked decline (p. 77). While 18% of people they surveyed were born on farms, only 6.8% were still located on farms by 1994, and it is likely that this number has reduced since then. Most residents of Carnarvon have had a significant historical link with the town, and a long historical link with the Northern Cape. Indeed, 68% of people in Kareeberg Local Municipality have never moved. The study points out that while a trend of declining employment is not different from elsewhere in South Africa, what is noteworthy about small Karoo towns such as Carnarvon is that there is a limit to which job losses in one sector can be absorbed by increases in another. In Carnarvon, half of the people employed are classified in 'Elementary' occupations' (p. 74).

All of the government reports and development planning documents recognise high levels of poverty, unemployment, and dependence on social grants, particularly the age pension. They also recognise problems of alcoholism and substance abuse, domestic violence, teenage pregnancy, and increasing numbers of children with the symptoms of foetal alcohol syndrome.

Concluding comments

This chapter has focused on key features of the political transition that ended apartheid and brought constitutional democracy into being in South Africa. It was this context of political compromise that ended the formalities of colonialism, but in ways that enabled deep structural inequalities to linger on, albeit often in new forms.

Sharpening the focus to the vantage point of Carnarvon, the chapter has shown how different provincial and local government administrations, together with a different approach to spatial development planning, have reconfigured the representations of space as well as the possibilities and limits of change. In Carnarvon, social grants have served to ameliorate existing poverty, but patterns of ownership of land and economic resources have not shifted, and opportunities remain constrained in what is defined as a marginalised part of the country in development terms. Although apartheid assumptions about segregation have been disrupted, they have by no means disappeared. Change has been uneven, with uncertain consequences for building a shared moral order.

The detail of this chapter provides the context for the analysis of schooling in the following chapters, and in particular the contestations that took place when new education policies challenged the segregated and unequal schooling that was apartheid's legacy. What emerges more clearly in the following chapters is that the changes of the post-apartheid period were not sufficiently radical to shake a sense of white entitlement or to build a new social imaginary for a shared community of equal participants.

Notes

1 The ANC, South African Communist Party (SACP), and Congress of South African Trade Unions (COSATU) formed the Tripartite Alliance, which has held up until the time of writing (the 2019 elections).

2 Though the ANC had initially condemned the establishment of Bantustans, in the later days of the liberation it forged links with traditional leaders to gain a foothold in rural areas, and despite some ANC leaders viewing traditional leadership as anachronistic and had no place under democratic rule, a mutually convenient relationship was established between the ANC, looking for rural support and traditional leaders seeking to shore up their powers. Influential in giving voice to traditional leaders was the situation in KwaZulu, a powerful bantustan that had refused independence under the leadership of the Inkatha Freedom Party and Gatsha Buthelezi, with an established king who demanded recognition. Contralesa, the Council of Traditional Leaders in South Africa, had been formed in 1987. (See Baldwin, 2015; Oomen, 2005; Williams, 2004, 2010.) The Traditional Leadership and Governance Framework Act of 2003, the Communal Land Rights Act (CLRA) of 2004, and the Traditional Courts Bill (TCB) of 2011in particular have extended the reach of customary law.

3 There is extensive scholarship showing the depth, severity, and persistence of poverty, inequality, and unemployment over time in post-apartheid South Africa, including in particular the work carried out by members of Southern African Labour and Development Research Unit (SALDRU) at the University of Cape Town, led by Murray Leibbrandt. Among other work, their studies of successive waves of the National Income Dynamics Survey are important in illustrating trends and dynamics over time.

4 These authors delineate five income categories in the South African population: chronically poor, 48.79%; transient poor (moving in and out of poverty), 11.5%; vulnerable middle class, 15.9%; stable middle class, 20.75%; elite, 3.6%.

5 Branson et al. (2012) sum up the situation as follows: 'Low school quality results in students completing fewer years of school, having a lower chance of employment when entering the labour market, and experiencing lower returns to education when employed, than do students who attended higher quality schools' (p. 14).

6 Legislation on local government includes the Municipal Structures Act (1998), the Demarcation Act (1998), the Municipal Systems Act (2000), and the Municipal Finance Management Act (2003).

7 There is an extensive literature on local government changes, which is not detailed in this book except for works specifically referenced.
8 The extent of local level protests and the significance of social movements in the post-apartheid period have been well documented. (See Ballard, Habib & Valodia, 2006; Bond, 2014; Gibson, 2006.) For an analysis of complex citizenship see Chipkin (2003); for elite corruption, von Holdt (2013).
9 See also the exchange between Pointer (2004) and Desai and Pithouse (2004) in the *Journal of Southern African Studies*.
10 The National Party withdrew from GNU in 1996 and formed the New National Party. The NNP merged with the Democratic Party in 2000 to form the Democratic Alliance but withdrew from the DA in 2001. The Freedom Front was formed by General Constand Viljoen in 1994 and changed to the FF+ in 2003.
11 Interview with E. Reilly, former mayor – who also pointed out that the three towns were not satisfied with being brought together into one municipality. Carnarvon. 22 November 2019.
12 This was the most recent census at the time of writing this book. Though the general population profile is likely to be the same, it is likely that some information will have changed, particularly in terms of education levels and Internet access. It is also possible that unemployment will have increased.
13 Research Reports by the Centre for Development Research (2010) commissioned by the Square Kilometre Array (SKA).

References

Alexander, N. (2002). *An ordinary country: Issues in the transition from apartheid to democracy in South Africa*. Pietermaritzburg: University of Natal Press.

Alexander, P. (2010). Rebellion of the poor: South Africa's service delivery protests – A preliminary analysis. *Review of African Political Economy*, *37*(123), 25–40.

Alvaredo, F., Chancel, L., Piketty, T., Saez, E., & Zucman, G. (2018). *Global inequality report*. World Inequality Lab. Creative Commons Licence 4.0 - CC BY-NC-SA 4.0 World Inequality Lab, 2017.

Atkinson, D. (2007). *Arid areas report, Volume 1: District socio-economic profile and development plans*. Centre for Development Support: University of the Free State.

Atkinson, D. (2019). When stars collide: Competing development paradigms in the central Karoo. *Journal of Southern African Studies*, *45*(4), 689–709.

Baldwin, K. (2015). *The paradox of traditional chiefs in democratic Africa*. Cambridge: Cambridge University Press.

Ballard, R. (2005). Social movements in post-apartheid South Africa: An introduction. In S. P. Jones & K. Stokke (Eds.), *Democratising development: The politics of socio-economic rights in South Africa* (pp. 77–100). Leiden: Martinus Nijhoff Publishers.

Ballard, R., Habib A., & Valodia, I. (2006). *Voices of protest: Social movements in post-apartheid South Africa*. South Africa: University of KwaZulu-Natal Press.

Bhorat, H., & Goqa, S. (2013). The gender wage gap in post-apartheid South Africa: A re-examination. *Journal of African Economies*, *22*(5), 827–848.

Bond, P. (2014). *Elite transition: From apartheid to neoliberalism in South Africa* (Rev. and expanded edn.). London: Pluto Press.

Branson, N., Garlick, J., Lam, D., & Leibbrandt, M. (2012). *Education and inequality: The South African case*. A Southern Africa Labour and Development

Research Unit Working Paper Number 75. Cape Town: SALDRU, University of Cape Town.

Centre for Development Support. (2010). *Socio-economic follow-up study: Carnarvon and Williston*. Bloemfontein: University of Free State.

Chipkin, I. (2003). 'Functional' and 'dysfunctional' communities: The making of national citizens. *Journal of Southern African Studies, 29*(1), 63–82.

Claassens, A. (2014). Denying ownership and equal citizenship: Continuities in the State's use of law and 'custom', 1913–2013. *Journal of Southern African Studies, 40*(4), 761–779.

Corder, H. (1994). Towards a South African constitution. *The Modern Law Review, 57*(4), 491–533.

Department of Provincial and Constitutional Affairs. (1998). *White paper on local government*. Pretoria: Department of Provincial and Constitutional Affairs, Republic of South Africa.

de Visser, J. (2009). Developmental local government in South Africa: Institutional fault lines. *Commonwealth Journal of Local Governance, 2*(January), 7–25.

Desai, A., & Pithouse, R. (2004). Sanction all revolts: A reply to Rebecca Pointer. *Journal of Asian and African Studies, 39*(4), 295–314.

Friedman, S. (Ed.). (1994). *The small miracle: South Africa's negotiated settlement*. Johannesburg: Ravan Press.

Gibson, N. C. (Ed.). (2006). *Challenging hegemony: Social movements and the quest for a new humanism in post-apartheid South Africa*. Trenton, NJ: Africa World Press.

Goebel, A. (2011). Our struggle is for the full loaf: Protests, social welfare and gendered citizenship in South Africa. *Journal of Southern African Studies, 37*(2), 369–388.

Kareeberg Local Municipality. (2017). Integrated development plan review, 2017–2022. Carnarvon.

Marais, H. (1998). *South Africa limits to change: The political economy of transformation*. London: Zed Books Ltd.

Nel, de W. (2015). *Karoo kaleidoscope: A historic perspective*. Carnarvon: Self-published.

Nel, E., & Hill, T. (2008). Marginalisation and demographic change in the semi-arid Karoo, South Africa. *Journal of Arid Environments, 72*, 2264–2274.

Nel, E., & Rogerson, C. M. (2009). Re-thinking Spatial Inequalities in South Africa: Lessons from international experience. *Urban Forum, 20*, 141–155.

Northern Cape Province. (2012). *Northern Cape Provincial spatial development framework, Volume 2: Profile and Key Aspects*. Kimberley: Government of the Northern Cape Province.

Office of the Auditor General. (2017). *Consolidated general report on the Local Government Audit Outcomes, 2016–17*. Pretoria: Auditor General of South Africa.

Oomen, B. (2005). *Chiefs in South Africa: Law, power and culture in the post-apartheid era*. Oxford: Curry.

Picard, L., & Mogale, T. (2014). *The limits of democratic governance in South Africa*. Boulder, CO: Lynne Rienner.

Pointer, R. (2004). Questioning the representation of South Africa's 'new social movements': A case study of the Mandela Park anti-eviction campaign. *Journal of Asian and African Studies, 39*(4), 271–294.

Powell, D. (2012). Imperfect transition: Local government reform in South Africa 1994–2012. In S. Booysen (Ed.), *Local elections in South Africa: Parties, people, politics* (pp. 11–30). Bloemfontein: Sun Press.

Rogerson, C. (2011). Tracking local economic development policy and practice in South Africa, 1994–2009. *Urban Forum, 22*(2), 149–168.
Seekings, J., & Nattrass, N. (2015). *Policy, politics and poverty in South Africa.* London: Palgrave Macmillan.
Segal, L., & Cort, S. (2011). *One law, one nation: The making of the South African Constitution.* Johannesburg: Constitutional Hill Trust; Jacana.
Siddle, A., & Koelble, T. (2012). *The failure of decentralisation in South African local government: Complexity and unanticipated consequences.* Cape Town, South Africa: UCT Press.
South African Local Government Association (SALGA). (2015). *15 Years of developmental and democratic local government, 2000–2015.* SALGA.
Statistics South Africa. (2018). Who is most likely to be affected by long-term unemployment? http://www.statssa.gov.za/?p=11688. Posted 30 October 2018.
Statistics South Africa. (2019). Five facts about poverty in South Africa. http://www.statssa.gov.za/?p=12075. Posted 4 April 2019.
Strand, P. (2001). Finalizing the South African Constitution: The politics of the Constitutional Assembly. *Politikon: South African Journal of Political Studies, 28*(1), 47–63.
Terreblanche, S. (2011). *Lost in transformation: South Africa's search for a new future since 1986.* South Africa: University of KwaZulu Natal Press.
von Holdt, K. (2013). South Africa: The transition to violent democracy. *Review of African Political Economy, 40*(138), 589–604.
Williams, J. M. (2004). Leading from behind: Democratic consolidation and the chieftaincy in South Africa. *Journal of Modern African Studies, 41*(1), 113–136.
Williams, J. M. (2010). *Chieftaincy, the state, and democracy: Political legitimacy in post-apartheid South Africa.* Bloomington: Indiana University Press.
World Bank. (2018). *Overcoming poverty and inequality in South Africa: An assessment of drivers, constraints and opportunities.* Washington: World Bank.
Zizzamia, R., Schotte, S., & Leibbrandt, M. (2019). Snakes and ladders and loaded dice: Poverty dynamics and inequality in South Africa, 2008–2017. Southern Africa –Towards inclusive economic development (SA-TIED), Working Paper #33.

7 Preserving privilege in schooling

From the vantage point of Carnarvon

Carnarvon's schools

At opposite sides of the town of Carnarvon stand its two schools. On the northern side, in the former Coloured township of Bonteheuwel, are the banks of face brick classrooms of the primary school, once the secondary school that was purpose-built by Coloured Affairs on the southern side, near where the now dry spring of Harmsfontein once attracted settlers, are the extensive but older buildings of the former white school, currently the town's secondary school. [1] Both schools reflect their apartheid architecture and geography, though neither serves the purposes of its original design. Comparatively speaking, the serviceable Bonteheuwel buildings, designed for low maintenance, appear less in need of care than the grander buildings of the former *Hoërskool* Carnarvon with their painted façade, now a little shabby. Neither school has sporting facilities that are maintained, though, again, the two sets of premises give contrasting impressions. The stony and usually unmarked sports field of the primary school conveys a semblance of serviceability. The facilities surrounding the secondary school convey a powerful sense of neglect: the broken nets and crumbling surfaces of the tennis courts, the empty swimming pool, and the unmaintained show grounds suggest the passing of an era of more resources and higher maintenance.

Behind the buildings of these two schools lies a story that divides the town: the story of school restructuring from apartheid segregation to a single education system. At the time of change in the 1990s, the parent body of the white school refused to amalgamate with the Coloured schools and was able to use its legal powers under the new South African Schools Act to resist government instruction for ten years – even though the two Coloured schools in the town were clearly overcrowded on a single site. When the three schools were finally restructured into two in 2007, the anger of the white community resulted in the bitter withdrawal of its children and its resources from the town's schools. This chapter and the following one explore this story from two angles: first, how it was possible for the white parent body of *Hoërskool* Carnarvon to resist change in the post-apartheid schooling dispensation and second, the social imaginary of difference that their actions exhibit. Across both chapters is a story of decolonisation and the lingering inequalities of coloniality.

While Carnarvon's story is a specific one, looking at policy change from the vantage point of this particular place enables power relations within the post-apartheid education settlement to be magnified and illuminated. Carnarvon's story speaks to a specific instance on the larger landscape of schools and inequality – the unevenness of which I return to in the conclusion to this chapter.

This chapter begins by outlining the context of change in education as apartheid ended, and the establishment of new administrative structures and new legal framework for a single system of education for all. The chapter argues that changes in education were fundamentally shaped by decisions taken in the domain of constitutional negotiations and the battle between political interests, rather than the domain of the education policy process, and that policy implementation was thwarted by the agency of local interests fighting against it. I argue that agreements reached in constitutional negotiations provided space for former white schools to maintain a privileged position in the post-apartheid dispensation, carrying existing inequalities forward into the post-apartheid future and reconfiguring them. Decolonising schooling, I argue, needs to confront these structural inequalities and the social imaginary that justifies them.

Designing and implementing a new education system

In the end days of apartheid, it was clear to all across the political spectrum that schooling would need to change. Apartheid education, with its profound inequalities, racial discrimination, and spatial segmentation, was symbolic of the old order which needed to give way to the new. Even so, of all the changes tackled by the post-apartheid government, schooling was one of the hardest to reshape. Schooling is more complex as a social institution than it may appear to be from the familiar structures of schools and classrooms. Behind the visible forms of schools lie the multi-layered bureaucracies that administer them, and policies for schooling are influenced by power plays and vested interests of all sorts. Schools are part of the social imaginary as emblems of the future and they are imbued with personal as well as social aspirations. And schooling systems are more resilient to change than policy makers anticipate. These points will become evident as the post-apartheid education policy settlement is traced in this chapter, together with the narrative of how policy implementation was contested in Carnarvon's schools.

Much has been written about the struggle against apartheid education, and about the raft of education policies introduced by the post-apartheid government.[2] Alongside this existing scholarship, I weave together an account of the constitutional negotiations and the accompanying developments in education policy. In this sketch, I pay particular attention to the development of policies on school governance, since these policies emboldened *Hoërskool* Carnarvon to stand against the provincial government's instructions. I then look at what happened when new policies were introduced to restructure Carnarvon's schools, showing how local agency shaped implementation in favour of those with economic and social power.

Apartheid's dying days

In the heightened political struggle of the 1980s, schooling was a focal point. The Soweto student uprisings of June 1976 had spread across the country, with students boycotting schools and protesting against 'Bantu Education', labelled an inferior system of 'gutter education'. Student protests linked up with other mass mobilisations to render schools and townships ungovernable. At the peak of the education struggles, Parent Teacher Student Associations (PTSAs) were regarded as an alternative authority structure for schools, and the National Education Crisis Committee (NECC) was formed to focus the demands for change.[3] Calls for 'People's Education' gestured towards possibilities of a radically new system, based on equality, democracy, and non-discrimination.

At the same time, white education was under pressure of a different sort. Many white schools, which received privileged funding and resources under apartheid, were experiencing a drop in student numbers and some were closing. The under-utilisation of white schools stood in stark contrast to the overcrowding and under-resourcing of black schools. Funding disparities between the different apartheid education departments presented problems of a different order for the National Party (NP) government, since equalising all expenditure at the level of the minority white system was out of reach for the budget. Under these circumstances, a major concern for the government was how to secure the future of white schools – Afrikaans-medium schools in particular – given that change was inevitable.

Through the so-called 'Clase Models', the NP government devolved significant powers to the parent bodies of white schools and permitted them to admit students of other races under strict conditions. Schools were given three models to choose between: becoming a fully private school (Model A); remaining a state school (Model B); or becoming a state-aided school, having responsibility for raising a portion of their own funding (Model C). Very few schools chose Model C, most preferring to stay as they were. In 1992, however, the government declared all white schools to be Model C. This meant that at the time of political change, all white schools were changed into state-aided public schools with white-controlled management councils having responsibility for raising part of their funding themselves. However, regulations stipulated that these schools would continue to fall under white education departments, would remain majority white in their enrolments, and must preserve their 'traditional values and ethos' (Christie, 1995; Karlsson, 2002). Through these measures, Model C schools were able to preserve their existing profile and ethos as white schools with tightly controlled desegregation, and they were ring-fenced in the negotiation process (as I shall show). *Hoërskool* Carnarvon was one such Model C school.

Negotiating a new education policy framework, 1994–1996

Constitutional negotiations

In the tough context of constitutional negotiations discussed in Chapter 6, important agreements were reached on education (see Segal & Cort, 2011). These had defining implications for policies on schooling that were being developed in a separate process altogether. Important constitutional negotiations related to education as a basic right, the status of existing school governing bodies (SGBs), and language rights in public schools. Of these, language rights in school became the most contentious point. It was only after negotiations on the final constitution were concluded in 1996, that the final versions of the major laws restructuring education – the National Education Policy Act (NEPA) and the South African Schools Act (SASA) – were gazetted. In other words, new policies for schooling reflected not only education policy deliberations but also the broader negotiations of this period, with key negotiating points being governance and language rights in public schooling. These were points relating to the legal status of Model C schools and the protection of Afrikaans-medium public schools (as I shall show).

In the broader constitutional negotiations, the Interim Constitution of 1993 made the following provisions for language and education:

> 32 Education
> Every person shall have the right–
>
> a to basic education and to equal access to educational institutions;
> b to instruction in the language of his or her choice where this is reasonably practicable; and
> c to establish, where practicable, educational institutions based on a common culture, language or religion, provided that there shall be no discrimination on the ground of race.
>
> (1993, s. 32)

Explicit in this formulation was that single medium instruction could be offered by private schools, but that this might not be 'reasonably practical' for publicly funded institutions. However, with regard to the status of Model C schools, article 247 of the Interim Constitution was of great importance:

> 247 Special provisions regarding existing educational institutions (1) The national government and the provincial governments as provided for in this Constitution shall not alter the rights, powers and functions of the governing bodies, management councils or similar authorities of

departmental, community-managed or state-aided primary or secondary schools under laws existing immediately before the commencement of this Constitution unless an agreement resulting from bona fide negotiation has been reached with such bodies and reasonable notice of any proposed alteration has been given.

(1993, s. 247)

Education policy actors

Between the elections of 1994 and the adoption of final constitution in 1996 – the period of the Government of National Unity (GNU) – new policies for education were developed. In Mandela's cabinet in the GNU, the new Minister of Education, Sibusiso Bhengu, had recently returned from exile and was from the higher education sector, as was the new Director General. They were not a strong team to restructure schooling, particularly given that the Deputy Minister, Renier Schoeman, was an experienced member of the NP and a number of senior officials from the apartheid period had remained in office in the new Department of Education. Very few ANC appointees to the Education Department had expertise in the schooling sector, let alone how to administer the system, not least because schooling had been so tightly controlled by the apartheid government.

Straddling competing interests and complex power plays, the new ministry adopted a cautious approach to government, using established rules and procedures of parliamentary governance to demonstrate its legitimacy. The governmentality of a modern state meant the official discourse of white papers, national commissions, portfolio committees, appointments to offices with line functions, and so on (see Christie, 2006). These conditions inevitably meant a disconnection from the dynamics that had energised the earlier protest movements, which had been responding to actual contexts and conditions on the ground (Chisholm & Fuller, 1996; de Clercq, 1997).

Rearranging powers and functions in education

The major task facing the new Ministry and Department of Education was to set up a new and unified system of schooling, while still running the existing system. In Foucauldian terms, its first task was to draft regulations for running itself – 'the conduct of conduct'. The racially divided and fragmented education departments (19 of them) were dissolved, and nine new provincial departments were set up alongside a single new national department.[4] In line with the interim constitution, the new national department of education was given responsibility for drawing up 'norms and standards' and developing framework policies for the system as a whole, while provincial departments were given responsibility for implementing national policies, delivering services, and providing schools.

White Paper 1 and Model C schools

In 1995, after much wrangling within the Department, the first *White Paper on Education and Training* was released, outlining the values and principles for the new system of education of equal quality for all, based on the 1993 interim constitution. Straddling different political interests, the White Paper had two introductions, one from ANC Minister Bengu and a second from NP Deputy Minister Schoeman. Bengu's optimistic introduction emphasised the need to build a common system based on equity and non-discrimination, as well as the effective use of all resources. Schoeman's introduction, cautious in tone, noted that there was a difficult road to travel to reach 'a truly national consensus' on the way forward.

While specifically grounding itself in the principles of the new constitutional order, the White Paper also signalled that agreements reached in constitutional negotiations would need to be respected. Notably, it referred to the agreement brokered by the NP in constitutional negotiations that the existing status of Model C schools would be respected:

> The national and provincial governments are required to reach agreement 'by bona fide negotiations' with the respective governing bodies, and give reasonable notice, before altering the rights, powers and functions of such bodies. If agreement is not reached by negotiation, a government may nevertheless proceed to make the alterations it wishes. If it does so, however, the Constitution gives 'interested persons or bodies' a specific entitlement to mount a legal challenge to the validity of such alterations in terms of the Constitution.
>
> (1995a, p. 679)

In effect, this meant that Model C schools were ring-fenced and protected, and that they would have the right to legally challenge attempts to change them. This was very significant matter for policies developed to address the governance and funding of schools, as will be seen.

The White Paper placed on record the deep structural inequalities between schools, outlining these as follows:

> 19. It is understandable that many parents, school principals, teachers and students are uncertain about what the changes in the system of education will mean for their schools and themselves. Those who are accustomed to stable schools, which have close links with the social, cultural and religious life of their communities, and honoured traditions, may feel that what is precious to them is threatened by unknown changes they may be unable to influence or control. Communities which have been favoured by the past political dispensation, and who know that a democratically elected government, representing an overwhelmingly poor electorate, cannot be expected to fund their privileges, may be particularly apprehensive about what is in store.

> 20. Equally, parents, teachers and students who have had to cope with appalling conditions, the result of decades of under-resourcing, instability, wasted human potential and low morale, have high expectations from a government they believe rightly is committed to redress.
>
> (1995a, p. 69)

This comparison certainly applied to the schools in Carnarvon. *Hoërskool* Carnarvon was a 'stable school' with 'close links to the social, cultural and religious life' of the white Afrikaner community. Its parent body was determined to protect what they regarded as 'precious to them ... threatened by unknown changes'. On the other side of town, the two Coloured schools, cramped onto one site, were 'under-resourced' and though their operating conditions were not as 'appalling' as those of many black rural and township schools, they had certainly faced 'decades of under-resourcing' and at times, 'low morale'. These two schools would certainly have been justified in having 'high expectations from a government' that was 'committed to redress'.

In recording the deep structural inequalities in the schooling system and committing itself to change, the White Paper was unequivocal about the need for change:

> The present pattern ... is a patchwork from the past. It contravenes the rights to equality and non-discrimination which the Constitution guarantees. As a basis for a national system in a democratic South Africa it is dysfunctional and cannot continue unchanged.
>
> (1995a, p. 68)

In light of this, it recommended that a review committee be set up to explore alternatives for governance and funding.

On the basis of the White Paper, the department set to work on designing an umbrella framework of developmental initiatives and proposals. In brief summary, the new policy suite for schooling is set out in Table 7.1. This framework of ideal-type policies – drawn up without proper consideration of provincial capacity or actually existing conditions in schools – has been bedevilled by problems of implementation from the start.

Debates on school governance and funding

To canvass possibilities for governance and funding, a review committee was established (the Hunter Committee) to propose a framework that would be 'likely to command the widest possible public support', would 'improve the quality and effectiveness of schools and [would] be financially sustainable from the public funds' (1995b, p. 68)

The recommendations of the Committee, published in August 1995, were distilled into White Paper Two on the Organisation, Governance, and

Table 7.1 Post-apartheid education policies

National Education Policy Act (1996)
- Amended apartheid laws to dismantle racially separate education departments.
- Established new departments and allocated responsibilities to the national department for setting norms and standards for the system, and to provinces for the implementation of schooling, the employment of teachers, and other matters, in accordance with national policy.

South African Schools Act (SASA) (1996)
- Provided new arrangements for the governance and funding of schools.
- Set out the rights and responsibilities of school governing bodies (SGBs) as juristic persons.
- Allowed schools to charge fees, within certain guidelines.
- Enabled SGBs to set policies on a range of matters including admissions and language of instruction, provided these were not used to discriminate on racial grounds.

National Norms and Standards for School Funding (1998)
- Divided schools across the country into quintiles from least poor (Q5) to poorest (Q1).
- Provided for equity funding from a small part of the education budget to give more funding to schools in poorer quintiles.
- Later, Q1–3 were declared fee-free.

Education Labour Relations Act (1995)
- Set out frameworks for equal conditions of employment for teachers, including conditions of work, codes of conduct, and duties and responsibilities were agreed upon for educators.
- All teachers were required to register with the South African Council of Education (SACE).

Curriculum framework
- Curriculum 2005 was introduced in 1997 to replace the apartheid curricula with a national Outcomes Based Education (OBE) curriculum framework; this gave very little guidance on curriculum content.
- Revised National Curriculum Statements (RNCS) of 2002 and National Curriculum Framework (NCF) of 2007 retained the OBE curriculum design but provided more detail on content and assessment.
- Curriculum and Assessment Policy Statements (CAPS) of 2012 revised the NCS to abandon OBE and provide stronger guidelines for content and pacing.
- Progression policies stipulated that students who did not achieve the expected standard to pass the grade (i.e. who failed) could be held back only once per phase, and would then be automatically promoted to the next grade. (The phases are: Foundation Phase, Grades 0–3; Intermediate Phase, Grades 4–6; Senior Phase, Grades 7–9; Further Education and Training, Grades 10–12). This meant that all classes could include students who had not achieved the expected standard for the grade.

Language in Education Policy (1997)
- This policy was developed separately from curriculum policies and promotes 'additive multilingualism' and English assimilationism

Special Needs
- A National Commission on Special needs in Education and Training: this advocated 'mainstreaming' of learners with special educational needs and was followed by White Paper Six on Inclusive Education (2001).

Early Childhood
- White Paper on Early Childhood Development (2000) envisaged the introduction of a reception grade (Grade R).

Funding of Schools (February 1996) and the South African Schools Bill (April 1996). The shared assumption across the political spectrum was that school governance would be decentralised in one form or another – though this reflected different political aspirations (see Beckmann & Prinsloo, 2015; Carrim, 2001; Clase, Kok & van der Merwe, 2007; Karlsson, 2002; Karlsson, McPherson & Pampallis, 2001; Lewis & Naidoo, 2006; Prinsloo, 2006; Sayed, 2010; Tikly & Mabogoane, 1997). From the side of the mass democratic movement, SGBs were viewed as a form of democratic participation to restore legitimacy to schools which had been disrupted by protests against apartheid education. For conservative white groups, SGBs would be able to protect the historically privileged schools at the time of change. However, as Frances Eberhard (2016) points out:

> What emerges from a look at the drafting history of the Schools Act is that the approach to school governance arrived at represents *both* a manifestation of the ANC's Redistribution and Development Programme's push towards participative democracy, *and* a way to diffuse tensions around retention of privilege in formerly white schools.
>
> While both of these motivations can co-exist at the point of drafting the legislation, tensions arrive in implementation. Retention of minority power over formerly white schools requires no action other than a hands-off approach on the part of the State. True democratisation of schools, on the other hand, requires the traditional power balances to be altered.
>
> (p. 4)[5]

Whereas governance was a point of agreement, options for funding were more complex given the apartheid disparities and power imbalances. The Hunter Committee had proposed three alternatives for consideration, but in the process of discussions and negotiations on the Schools Bill none of these was selected. Instead, at the final stages of discussion, the government accepted the advice of international consultants that schools be given wide powers to charge fees in addition to state funding. Their advice was based on the argument that this would keep white parents, and a nascent black middle class, within the public schooling system – and this measure would also increase the funding available for the schooling system. The advice also reflected the global trend of marketisation, offered in the South African context as a policy compromise that would protect existing white interests. However, this policy would have fundamental consequences for any equity agenda, confirming the existing resource inequality and allowing it to grow.

Final Constitution, 1996

In the broader constitutional negotiations (which were concluded in October 1996), agreements were reached on many difficult issues, but three clauses remained to be resolved as the deadline drew close: the clauses on property rights, labour lockout, and education. In education, the NP and ANC were deadlocked on the issue of single-language institutions (Segal & Cort, 2011). Both agreed that private schools had the right to be single-language institutions, but the ANC was not prepared to entrench this right in state-financed schools. Finally, at the eleventh hour, an agreement was reached based on the Democratic Party's proposed amendment that single medium schools would 'not be precluded' from receiving state subsidies.

Given the plethora of matters discussed at this time, and the many education policy proposals that were on the table, it is easy to overlook the importance of this particular constitutional detail in determining the policy arrangements that were to follow. The fact that reaching this compromise on education was crucial to the adoption of the whole constitution is a point that is seldom highlighted in the education literature of this period.

The Constitution was adopted in October 1996, with the final education clauses reading as follows:

> Education
>
> 29
>
> 1 Everyone has the right –
>   a To a basic education, including adult basic education; and
>   b To further education, which the state, through reasonable measures, must make progressively available and accessible.
>
> 2 Everyone has the right to receive an education in the official language or languages of their choice in public educational institutions where that education is reasonably practicable. In order to ensure the effective access to, and implementation of, this right, the state must consider all reasonable educational alternatives, including single medium institutions, taking into account –
>   a equity;
>   b practicability; and
>   c the need to redress the results of past racially discriminatory laws and practices
>
> 3 Everyone has the right to establish and maintain, at their own expense, independent education institutions that –
>   a do not discriminate on the basis of race;
>   b are registered with the state; and
>   c are not inferior to standards at comparable public educational institutions.
>
> 4 Subsection (3) does not preclude state subsidies for independent educational institutions.
>
> (1996, s. 29)

The National Education Policy Act (NEPA) was gazetted in April 1996, and once the Constitution had been adopted in October 1996, the final version of the South African Schools Act (SASA) was passed in November.[6] In other words, SASA could not be finalised until the new constitution was approved – a point that could be overlooked in the multiplicity of contested policy developments of this period.

South African Schools Act (1996)

At the end of 1996, the South African Schools Act was finally passed. In its preamble, SASA set out its vision for a new schooling system, on the basis that 'the achievement of democracy in South Africa has consigned to history the past system of education which was based on racial inequality and segregation'. The preamble envisaged:

> a new national system for schools which will redress past injustices in educational provision, provide an education of progressively high quality for all learners and in so doing lay a strong foundation for the development of all our people's talents and capabilities, advance the democratic transformation of society, combat racism and sexism and all other forms of unfair discrimination and intolerance, contribute to the eradication of poverty and the economic well-being of society, protect and advance our diverse cultures and languages, uphold the rights of all learners, parents and educators, and promote their acceptance of responsibility for the organisation, governance and funding of schools in partnership with the State. (Preamble)

SASA gave extensive powers to SGBs, going further than the powers that Model C schools had been given. While they were bound by constitutional requirements of equal rights and non-discrimination, SGBs were able to set their own policies on admissions, language, religion, and code of conduct for staff and students. They had extensive powers over selecting staff and charging fees, and most had financial powers. Significantly, they had the legal status of juristic bodies, and were able to contest provincial and national departments in the courts if they believed their rights were infringed. The result has been a significant amount of litigation on education – in which *Hoërskool* Carnarvon was one of many participants.

Resistance to restructuring in Carnarvon

Carnarvon's schools in 1996

At the time of negotiations, Carnarvon's three schools, all with hostel facilities to serve the farming community, had fallen under racially based departments of education. The white school, *Hoërskool* Carnarvon, was a Model C school under what was then the Cape Provincial Education Department, and as a 'combined school', it provided for all 12 grades (Grade One to Standard Ten). The town's two Coloured schools were controlled by the 'own affairs' Coloured House of Representatives and funded at a lower level. The primary school, *Laerskool* Carel van Zyl, the original Rhenish Mission School established in 1848, was the only school available for Coloured residents until Carnarvon *Sekondêreskool* was purpose-built in Bonteheuwel in 1984. As detailed in Chapter 5, the *Laerskool* had been relocated to share the premises of the *Sekondêreskool* in Bonteheuwel after it burnt down in 1989. Clearly this was not envisaged as a permanent measure, since a site had been earmarked for the construction of a new school near what was known as '*Sekondêr*', which was already full.

The result of this relocation was less than satisfactory for both schools. It meant that there were two separate schools working from a single set of premises that had been purpose built for the secondary school. Two timetables were operating, two sets of bells ringing, two sets of noisy break times, and extreme overcrowding of all facilities. Toilet facilities, in particular, were inadequate for the number of students, leading to a crisis in sanitation as sewage overflowed into the street.

At the same time, on the other side of town, *Hoërskool* Carnarvon continued to operate as a state-aided Model C school, tightly run by a parent management committee that had access to resources from the town and its farms, and was able to charge fees as deemed appropriate. It supplemented its numbers by admitting a select number of fee-paying Coloured students from the town. And in 1996 it appointed its first Coloured teacher, who had come across from *Sekondêr*. The white parent community invested in the school both financially and emotionally. The school took pride in its hostels and the appointment of *koshuismoeders en vaders* (hostel mothers and fathers), and parents contributed to the maintenance of the buildings and grounds, often on a voluntary basis. Some of the wealthier white parents continued an established tradition of sending older children to boarding schools in larger towns (including prestigious public schools that they themselves had attended), and some farmers had bought houses in the town where their children stayed. There is no doubt that the parents of *Hoërskool* Carnarvon, who included farmers as well as townspeople, had a strong sense of ownership, control, and investment in the school.

Legal contestation

Once SASA had been gazetted, the Education Department of the Northern Cape Province gazetted its own Schools Education Act in 1996 to incorporate SASA and implement its provisions for the start of the new school year in 1997. The member of the executive council (MEC) responsible for the Northern Cape Education Department (NCED) was the (Coloured) ANC member Tina Joemat.[7]

In a provincial notice in early 1997, MEC Joemat outlined steps to be taken to provide an integrated public school system for the province. Given its geographical spread (30% of South Africa's land mass) and small population (2% of the country's total) the Northern Cape faced particular challenges on how to stretch resources equitably across schools which had previously been part of three separate departments and funded differentially. MEC Joemat proposed school restructuring so that where practically possible, each rural town would have a junior primary school, a senior primary school, and a secondary school.[8] An implication of this was that the former Model C schools in the region – almost exclusively Afrikaans-medium schools – would be restructured. Confident in their constitutional rights to language and self-governance as enshrined in SASA, *Hoërskool* Carnarvon and the neighbouring schools in the region were determined to fight what they saw as unwelcome changes forced on them by a government that was unsympathetic to their interests.

As an indication of how Model C schools in the region viewed the changes that began in 1997, the articles, letters, and editorials of the Calvinia-based regional newspaper, *Die Noordwester*, provide interesting insights.[9] They show how the contestation over the schools in the region unfolded, and also the expectations of SGBs in relation to their rights in the face of the changes proposed by the NCED. Interesting, also, is the tone of the writers and their emphasis on process in terms of 'manners' and 'etiquette'. In short, these excerpts illustrate how the different expectations of the Department and former Model C schools led to a hardening of positions on both sides.

- In June 1997, the NCED issued a press release announcing SGB elections, and in response, a lead article in *Die Noordwester* (13 June) noted sharply that certain schools had had governing bodies in place for many years – a warning that these SGBs expected their status to be recognised and *bona fide* negotiations to be held.
- In July, the *Noordwester* carried a letter under the heading 'Joemat speaks of amalgamation: governing bodies left in the dark'. Describing a visit of the MEC and others to Calvinia and its schools, the writer complained as follows:

  > What I particularly object to is the manner in which Ms Joemat made this [amalgamation] known. Governing bodies are totally ignored; in fact, we heard this from members of staff and on the streets. Is this how governing bodies will be treated in the future? ... I think it is extremely unprofessional and insensitive for an MEC of Education to handle such a far-reaching and extremely sensitive matter in such a manner. (18 July 1997)

- A week later, the Chairperson of the SGB of *Laerskool* Brandvlei complained that schools were hearing about amalgamation second-hand from other schools, under the heading 'Minister must clean up her etiquette: Amalgamation forced on schools'. The writer asked '*Where is the professional etiquette of actions? Where is democracy*?' He continued:

  > Please clean up your etiquette and make a thorough analysis of whether amalgamation is a practical possibility; and look us (the

> community) in the eye and do this in a manner that is respectful. If you don't do this, you are busy sticking a very thick stick into a hornets' nest. Our communities live together in harmony and we will not permit people from outside to force amalgamation autocratically on us. (25 July 1997)

- At the beginning of August, the Chairperson of the SGB of *Hoërskool* Fraserburg wrote in support of previous letter writers, saying

  > How on earth can the Education Department of the Northern Cape expect the cooperation of the governing bodies of schools when they make shocking announcements at schools and in some cases simply on the street, without informing governing bodies?... According to the constitution, parents have the right to place their children in the school of their choice. How can this happen when the department makes one-sided announcements that bypass the legitimately elected governing bodies. Why then were governing body elections held in schools? Must we go back to farm schools and church schools where parents could decide for themselves where children would go for their education? Are parents now to be forced on the one hand to pay taxes (among other things for state schools) and on the other hand have to pay themselves for the education of their children? (1 August 1997)

- The following week, 8 August 1997, amalgamation was again frontpage news in the *Noordwester* under the heading '*Amalgamation: parents and teachers concerned*'. The article reported on a meeting of parents and teachers from Sutherland, Fraserburg, Carnarvon, Williston, Brandvlei, Calvinia, and Loeriesfontein to address urgent matters of school integration. The meeting was addressed by representatives of the national teachers' organisation, *Suid-Afrikaanse Onderwysersvereeniging* (SAOU), and the chair of the organisation's regional branch (*Noord-Kaapse Vereeniging van Beheerliggame*) in attendance.[10] The meeting resolved to set up a task team to meet with MEC Joemat to establish whether rural areas only were being singled out for amalgamation, and how she intended to proceed with the project. '*Furthermore, the question is: on which article in the act is she proceeding and what are the actual reasons for her steps*'. Joemat's motives were questioned as being based in politics not education, and the meeting resolved that schools would not regard visits by circuit managers as constituting formal negotiations.
- A September edition of the *Noordwester* featured the task team's report-back to schools in Carnarvon, Williston, Brandvlei, Calvinia, Loeriesfontein, Kenhardt, and Sutherland. MEC Joemat had declared her support for a central task team from the region to discuss her continuing plans for amalgamation. (12 September 1997).
- On 3 October, the *Noordwester* provided a report back from the discussions with the MEC under the heading 'Task team visits Joemat'.[11] MEC Joemat informed the meeting about the explanation she had given

for amalgamations in the Provincial Gazette of 14 February. In summary, national budget calculations were based on student numbers, which did not take into account the unique circumstances of the Northern Cape, with its wide geographical spread and sparse population. The result was that the Northern Cape would exceed its budget for 1997 by R180 million. Given that the Education Department was responsible for providing teaching facilities for all students from Grade 0 to Grade 9, it was necessary that all teaching facilities and buildings were utilised to the maximum. MEC Joemat emphasised that her reasons for restructuring were driven by economic and not political considerations and agreed that a team of investigators from the Department would visit each town to discuss the practicability of amalgamations. She also emphasised that she did not wish to impose on communities from above, and that the process would preferably be settled by negotiation, while mediation and arbitration could also be used. Ultimately, if an agreement could not be reached, parties could approach the courts. The task team reported that it was satisfied that it had achieved its goal.

- However, when the Department took steps towards the amalgamation of schools in Richmond, the *Noordwester's* editorial took a strongly oppositional stand:

    > The racist Kimberley regime of the ANC is precipitating the province into administrative chaos and nightmare ... The whole amalgamation attempt is being forced from on high on vulnerable parents and children who have no other alternative but to call on the law. Recently elected school governing bodies, which were announced with great fanfare by the ANC authorities, are being completely ignored in the decision-making processes that affect their schools and have to learn from teachers and learners about the harassment visits, announcements and threats from the Minister of the Executive Council for Education or her representative. Only teachers who carry membership cards of the ANC or one of its unions are listened to. (17 October 1997)[12]

- In December 1997, the local schools in Carnarvon apparently reached an agreement that there could be two schools in the town, one from Grade 1 to 12 on the premises of the *Hoërskool* and one from Grade 1 to Grade 7 on the premises of the *Sekondêreskool.* (This arrangement would have enabled segregation to continue at primary level.) However, the NCED rejected this proposal out of hand, replying by fax to the principals of the schools that it intended to implement its model of one high school and one primary school in the town from the start of 1998. It instructed schools to open in February instead of January in 1998 to allow for the changes. In response, the *Hoërskool* appointed legal representation to oppose the process, and opened in January in defiance of the directive – while the Bonteheuwel schools obeyed and delayed opening. Representatives of the SGB of *Hoërskool* Carnarvon appealed for donations for legal costs, citing the need to protect their 98-year legacy and preserve quality education.

- On 27 March 1998, the *Noordwester* reported that the matter of Carnarvon's schools had been settled out of court, and that the *Hoërskool* school would not be amalgamated or closed.
- In October 1998 the school was in legal battle with the Department over the payment of two teachers, and once again the Department lost its case in court, this time with a fine for not appearing and not serving necessary documentation. In an article entitled 'Tina loses another Court case' the *Noordwester* lamented the incompetence (*onbevoegdheid*) and lack of professionalism in the Education Department, noting with indignation that during the Budget Debate, the MEC had instructed her officials to ignore certain correspondence from governing bodies.

It is very significant to note that courts upheld the claims of *Hoërskool* Carnarvon on technical matters of the law, even while recognising that the inequalities between the schools were against the spirit of government policy. The following picture of the schools in Carnarvon is provided in the records of a court case pursued by the white *Hoërskool* Carnarvon in 1999:

> The learners of the school are predominantly, if not exclusively, from the socalled 'white' community of Carnarvon. The other two schools in Carnarvon share the same facilities, and their learners are almost exclusively from the socalled 'coloured' or 'black' communities of Carnarvon.
>
> The school has a capacity for 540 learners, whereas its actual number of learners is 250, being slightly less than one half of its capacity. The other two schools share the same facilities and are stretched to capacity.
>
> The medium of education in all three schools is the Afrikaans language and the schools are located within a radius of approximately two kilometres from one another. [13]

Hoërskool Carnarvon and other schools used every possible technicality in law to oppose restructuring, including the fact that SASA had made no specific provisions for 'combined schools' to be restructured. Legal cases between public schools and education departments were a new feature on South Africa's education policy terrain. Whereas under apartheid, education law was primarily a legislative and executive matter involving statutes and regulations, the new policy arrangements in education resulted in the production of a significant amount of case law (see Deacon, Colditz, Mellet & van der Merwe, 2016; Woolman & Fleisch, 2009). A consequence of this is that technical matters have prevailed over guiding principles on issues under dispute. This is succinctly noted in a constitutional court case by the presiding judge, himself a former anti-apartheid activist. Arguing for the distinction between 'the determination of guiding policy on the one hand, and its translation into legally binding enactments on the other', Constitutional Court Judge Albie Sachs stated:

> I prefer to begin by stating the obvious, namely that laws, regulations and rules are legislative instruments whereas policy determinations are not. As a matter of sound government, in order to bind the public, policy should normally be reflected in such instruments. Policy determinations cannot override, amend or be in conflict with laws (including subordinate legislation). Otherwise the separation between legislature and executive will disappear. [14]

Ironically, in the particular case referred to, it would appear from the written judgment that the Department of Education failed in its appeal because its case rested on a clause in SASA instead of a clause in NEPA. The salutary lesson is that when education policy is decided by courts, technical issues necessarily prevail over substantive debate or policy intentions.

Over the years, *Hoërskool* Carnarvon pursued litigation on various matters, and succeeded in staving off restructuring until 2007, explicitly opposing the stated intentions of government policy.[15] It used its rights under SASA to continue in a 'Model C' form to control admissions while operating at half of its capacity, charging fees, having staff appointments of its choice, maintaining its historical traditions and its definition of 'quality'. Whatever its reasons, the SGB of *Hoërskool* Carnarvon felt justified in serving a small selection of the town's young people and maintaining the control of the white community over 'its' school – even in the face of the untenable overcrowding in the other two public schools in the town. In this regard, it is worth referring to the justification for legal action given by the SGB of *Hoërskool*, appealing for donations in the *Noordwester* to assist with legal costs to fight against amalgamation at the end of 1997:

> The Governing body of *Hoërskool* Carnarvon regards this action as important for all platteland schools in the Northern Cape for whom quality education and the preservation of their own ethos is important.[16]

Clearly, sectional interests could prevail against broader concerns for building quality education for all in the local community.

In 2007, SASA along with NEPA was finally amended by government gazette to curb the powers of SGBs opposing government directives. The relevant amendment simply reads that SGBs 'must not in any manner interfere with or otherwise hamper the implementation of a decision made by the Member of the Executive Council or Head of Department in terms of any law or policy'.[17] This laid the legal basis for Carnarvon's schools to be restructured.

Amalgamation, 2007

In the intervening years from 1998 to 2007, while *Hoërskool* Carnarvon remained under the management of its white SGB, unsatisfactory overcrowding continued on the Bonteheuwel site, where *Laerskool* Carel van Zyl and

Carnarvon *Sekondêreskool* jostled with each other. I was told by several interviewees that the schools competed for control over the site, with the *Laerskool* claiming that it should prevail since it was much larger, and *Sekondêr* counter-claiming that it was there first, and the school had been purpose-built for it. At a time of retrenchments and 'right-sizing' of school staff across the country, several staff, including the principal of *Sekondêr*, had 'taken the package'. This had had destabilising effects, with five staff members being suspended over protests about the principal's replacement. It seems that frustrations in these two schools were directed inwards to competition and micropolitics, rather than outwards to challenge the indifference of *Hoërskool* Carnarvon and the inaction of the education department. Under continuously pressured circumstances, a number of teachers paid fees to send their children to *Hoërskool* Carnarvon (which, as a combined school, included both primary and secondary classes).

In 2007, a tipping point was reached for joint action by Coloured parents at *Hoërskool* Carnarvon. When an extra charge for textbooks was added to existing school fees, a group of Coloured parents decided to take a stand and have their opposition recognised. Protests led to the temporary suspension of the principal by the Education Department while the matter was being investigated. At this point, SGB elections were scheduled to take place across the province. Unbeknown to the existing SGB at *Hoërskool* Carnarvon, the group of Coloured parents decided to organise themselves as '*Besorgde Ouers*' ('Concerned Parents') to contest the SGB elections, and they campaigned in the Coloured community. To the utter surprise and shock of the white parent body, Coloured parents filled the hall on election night, and the *Besorgde Ouers* won the elections, thereby gaining control of the SGB.

With the SASA amendment of 2007 providing a legislative mandate, and a new SGB in place at *Hoërskool* Carnarvon (with no white parents on it), the Northern Cape Education Department commenced the restructuring of Carnarvon's schools. On a weekend in April, classes were moved from one school to the other. White teachers and students found themselves for the first time in the former Coloured school across town (now the town's primary school), while secondary teachers and students from the coloured school joined their counterparts in the newly formed high school on what had been the premises of *Hoërskool* Carnarvon.

Within two years of these changes, all of the white parents had removed their children from both schools and their hostels, although the white teachers in both schools remained more-or-less in place. White parents were in a position to choose between 'home schooling' their children in an informal support centre or sending them away to boarding schools in various larger towns. They pointed out that boarding school had always been an option for them; as one put it, 'The main difference is that whereas before they were half an hour away, now they're further away'. The school hostels currently cater to Coloured farm children, and neither is full.

In effect, the withdrawal of white parents meant that the schools in the town became working class schools – though the national quintile system of equity funding does not reflect this. The secondary school, with its former white premises, is classified as quintile 4, and therefore receives less equity funding than the primary, former coloured school, which is classified as quintile 1 – though both serve the same community. The primary school has been declared a 'no fee' school, and in practice, the secondary school operates as 'no fee' since it is not able to garner fees from its parent body. These two schools provide a snapshot of the schooling that the state is able to provide to its citizens where there is no parental supplementation. Their staffing allocation and class sizes are calculated in terms of provincial norms and standards, and the schools must make do with what they get (including class sizes of around 40 in the primary school). The buildings of both schools are in need of maintenance and sporting facilities are barely existent. There is almost no funding for out-of-school activities and schools must rely on external competitions such as Legoland to meet with other schools. There are many examples across the country of schools that operate with this minimum configuration of state provision, and the schools in Carnarvon may be understood as local variants of a broader pattern.

Ironically, the settlement reached in 2007 returned the Coloured schools to separate premises, as had been the case before the fire of 1989. The main difference was that the schools were now equal to others in a national system, with segregation officially over. Arguably, this arrangement represents a reduced version of SASA's dream of a new system of schooling that would combat racism, contribute to the eradication of poverty, and promote the well-being of all in society.

Conclusion and implications

This chapter has used the story of Carnarvon's schools from the end days of apartheid to current times as a vantage point to understand the policy arrangements for school governance that were put in place during the negotiated settlement that ended apartheid. It has shown the importance of locating the changes to schooling in the context of broader negotiations, rather than analysing them more narrowly as education policy. In tracing the policy trajectory on the governance of former white schools, the chapter argues that the powers given to SGBs were intended to ringfence them against change – and the case of *Hoërskool* Carnarvon shows how these powers could be used to preserve the special interests of former white schools.

The division of powers between education departments and schools set out in SASA emboldened schools such as *Hoërskool* Carnarvon to hold onto their privileged positions in direct contradiction to the policy intentions of the government – though not all former Model C schools have gone to such lengths. Certainly, the case of *Hoërskool* Carnarvon illustrates the slenderness (if not absence) of a shared vision for a new system of schooling in South Africa, where the notion of equal rights for all collides with historically embedded power and interests. In this case, the historical dynamic of exclusive possession trumped the vision of equal sharing, and the white community had the resources to choose the path of exiting above that of engaging with change.

The vantage point of Carnarvon's schools also points to the limits of what policies can achieve – whatever their values and aims – when those with resources and agency choose to go against them. Though policy templates have intended logics, these logics are easily overshadowed when policies are enacted in specific places where local actors with resources are able to bend policy intentions and justify this in terms of history and culture. The slippage between SASA's stated vision and the legislative remit of the Act is a classic case of the complexities of policy implementation. A consequence of this slippage – almost certainly unintended by the ANC majority in government – is that many decisions relating to the provision of schooling are placed in the hands of the courts rather than parliament.

The contestation over Carnarvon's schools provides a clear illustration of how the power relations of colonialism are able to linger on, even after governments have changed and new laws have been promulgated to express the dreams of a new social order based on equal rights and dignity for all. Longstanding patterns of possession and dispossession in this local place were shaken but not fundamentally shifted by the changes of 1994. It seems not too far-fetched to see the intensely politicised struggle for control over the schools as yet another move in the zero-sum game of possession/dispossession that had started more than hundred years earlier. Those with economic and social resources were able to push back against changes and – even to their own cost – find alternatives to preserve their preferences of exclusivity in schooling.

Carnarvon's story is an extreme one, but many former white schools remain places of privilege, supplementing state grants with parent fees, and controlling the admission of black students while preserving their historical ethos. Indeed, it is this historical ethos that has come to define 'excellence' in post-apartheid schooling. Race has made way for social class as the marker of distinction as these schools preserve their privileged relationship to the requirements of the national curriculum. In the terms set out by Teese and Polesel (2003), these schools are 'fortified sites', pooling the economic and social resources of parents to meet the demands of the curriculum, leaving schools without resources as 'exposed sites'. The story of *Hoërskool* Carnarvon shows a school changing from a 'fortified site' to an 'exposed site' as its parent and student body changed – the consequence of choices made by a community able to find alternative 'fortified sites' elsewhere for their children to attend, but unable to find it in themselves to share equally with those living next to them.

It is precisely lingering inequalities such as these – complex, intersectional power relations – that theories of de/coloniality bring to the fore. As outlined by Maldonado-Torres (2007), coloniality theorists address the long-standing patterns of power that survive the formal demise of colonialism:

> that define culture, labor, intersubjective relations, and knowledge production well beyond the strict limits of colonial administrations. Thus, coloniality survives colonialism. It is maintained alive in books, in the criteria for academic performance, in cultural patterns, in common sense, in the self-image of peoples, in aspirations of self, and so many other aspects of our modern experience. In a way, as modern subjects we breathe coloniality all the time and every day.
>
> (p. 243)

These lingering relationships of inequality are further discussed in the chapter that follows, which looks at the social imaginary that underpins them.

Notes

1 As mentioned in earlier chapters, Carnarvon's population is somewhat unusual in South African terms in that it does not include Africans in significant numbers. The Census of 2011 lists the population as 85.1% Coloured and 9.1% white. The vast majority (93.7%) were Afrikaans-speaking.
2 The substance of this chapter is informed by long engagement with an extensive literature on education in South Africa, an engagement which is too extensive to be cited as separate texts except where specifically referred to.
3 This grouping of students, parents, and civil society leaders was later renamed the National Education Coordinating Committee.
4 There are different ways to count the number of departments. At a national level and including bantustans, there were 15 departments, and in addition there were 4 white provincial departments, making a total of 19 overall.
5 Eberhard makes this point in the context of arguing that 'SGBs need to be adequately supported'.
6 Karlsson et al. (2001) provide helpful details on the policy process leading up to SASA.
7 Later Tina Joemat-Peterssen, and a member of the National Assembly.
8 For details on this, see Bennie Groenewald Primêre Skool en andere v Premier van die Noord-Kaap en 'n ander [1998] 3 All SA 426 (NC) 'Where the legislature gave a school the legal capacity to exercise its functions, such a school had locus standi to challenge in court any action through which such functions are changed or limited'.
9 Research interviews conducted in Carnarvon provide views in line with these, so I have decided to use the wording of newspaper articles rather than quote from interviews. The excerpts that follow are my own translations from Afrikaans.
10 Afrikaans-speaking organisations, translated as: South African Teachers' Union; Northern Cape Union of Governing Bodies.
11 The original Afrikaans ('Taakspan gaan kuier vir Joemat') has a somewhat informal connotation, perhaps disrespecting of her formal office.
12 The original heading reads 'So sê die redakteur: '*Rasisme Kelder N-Kaap Onderwys*'.
13 High School Carnarvon and another v MEC for Education, Training, Arts and Culture of the Northern Cape Provincial Government and another [1999] 4 All

SA 590 (NC). This case also outlines the provincial notice of 1997 where MEC Joemat set out possible arrangements for schools in rural towns.

14 Minister of Education v Doreen Harris [2001] 4 SA 1297 (CC) 11 BCLR 1157(CC), pp. 11–12.

15 See Deacon et al. (2016) for a general guide to significant cases; also Woolman and Fleisch (2009). The FEDSAS website also provides updated details.

16 Own translation. *Noordwester* 30 January 1998 'Ons veg vir gehalte onderwys in ons dorp' ('We fight for quality education in our town').

17 Education Laws Amendment Act (2007) s. 9b.

References

Beckmann, J., & Prinsloo, J. (2015). Some aspects of education litigation since 1994: Of hope, concern and despair. *South African Journal of Education*, *35*(1), 1–11.

Carrim, N. (2001). Democratic participation, decentralization and educational reform. In J. D. Jansen & Y. Sayed (Eds.), *Implementing education policies: The South African experience* (pp. 98–109). Cape Town, South Africa: UCT Press.

Chisholm, L., & Fuller, B. (1996). Remember people's education? Shifting alliances, state-building and South Africa's narrowing policy agenda. *Journal of Education Policy*, *6*(6), 693–716.

Christie, P. (1995). Transition tricks? Models for school desegregation in South Africa, 1990–93. *Journal of Education Policy*, *19*(1), 45–55.

Christie, P. (2006). Changing regimes: Governmentality and education policy in post-apartheid South Africa. *International Journal of Educational Development*, *26*, 373–381.

Clase, P., Kok, J., & van der Merwe, M. (2007). Tension between school governing bodies and education authorities in South Africa and proposed resolutions thereof. *South African Journal of Education*, *27*(2), 243–263.

de Clercq, F. (1997). Policy intervention and power shifts: An evaluation of South Africa's education restructuring policies. *Journal of Education Policy*, *12*(3), 127–146.

Deacon, J., Colditz, P., Mellet, S., & van der Merwe, J. (2016). *Case Law handbook on education: Twenty-five cases that helped shape Education Law*. Kenwyn: Juta.

Department of Education (DoE). (1995a). *White paper on education and training: Education and training in a democratic South Africa*. Pretoria: Department of Education, Republic of South Africa.

Department of Education (DoE). (1995b). *Report of the committee to review the organisation, governance and funding of schools*. (Hunter Committee.) Pretoria: Department of Education, Republic of South Africa.

Department of Education (DoE). (1996). *Education White Paper 2: The organisation, governance and funding of schools*. Pretoria: Department of Education, Republic of South Africa.

Department of Education (DoE). (1998). *National norms and standards for school funding*. Pretoria: The Department of Education, Republic of South Africa.

Eberhard, F. A. (2016). A preliminary review of school governing body support initiatives in South Africa. Unpublished Mimeograph. Cape Town: Norkitt Foundation.

Karlsson, J. (2002). The role of democratic governing bodies in South African schools. *Comparative Education*, *38*(3), 327–336.

Karlsson, J., McPherson, G., & Pampallis, J. (2001). A critical examination of the development of school governance policy and its implications for achieving equity. In E. Motala & J. Pampallis (Eds.), *Education and equity: The impact of state policies on South African education* (pp. 139–177). Sandown; South Africa: Heinemann.

Lewis, S. G., & Naidoo, J. (2006). School governance and the pursuit of democratic participation: Lessons from South Africa. *International Journal of Educational Development, 26*(4), 415–427.

Maldonado-Torres, N. (2007). On the coloniality of being: Contributions to the development of a concept. *Cultural Studies, 21*(2–3), 240–270.

Prinsloo, S. (2006). State interference in the governance of public schools. *South African Journal of Education, 26*(3), 355–368.

Province of the Northern Cape. (1996). *School Education Act, no. 6 of 1996*. Kimberley: Provincial Gazette Extraordinary, no 225, 4 April.

Republic of South Africa (RSA). (1994). *Constitution of the Republic of South Africa, Act 200 of 1993*. Pretoria: Government Printer.

Republic of South Africa (RSA). (1996a). *Constitution of the Republic of South Africa Act 108 of 1996*. Pretoria: Government Printer.

Republic of South Africa (RSA). (1996b). *South African Schools Act 27 of 1996*. Pretoria: Government Printer.

Republic of South Africa (RSA). (2007). *Education Laws Amendment Act 31 of 2007*. Pretoria: Government Printer.

Sayed, Y. (2010). Globalisation, educational governance and decentralisation: Promoting equity, increasing participation, and enhancing quality? *Compare: A Journal of Comparative and International Education, 40*(1), 59–62.

Segal, L., & Cort, S. (2011). *One law, one nation: The making of the South African constitution*. Constitutional Hill Trust. Johannesburg: Jacana.

Teese, R., & Polesel, J. (2003). *Undemocratic schooling: Equity and quality in mass secondary education in Australia*. Melbourne: Melbourne University Press.

Tikly, L., & Mabogoane, T. (1997). Marketisation as a strategy for desegregation and redress: The case of historically white schools in South Africa. *International Review of Education, 43*(2–3), 159–178.

Woolman, S., & Fleisch, B. (2009). *The Constitution in the classroom: Law and education in South Africa, 1994–2008*. Pretoria: Pretoria University Law Press.

South African court cases

Bennie Groenewald Primêre Skool en andere v Premier van die Noord-Kaap en 'n ander [1998] 3 All SA 426 (NC).

High School Carnarvon and another v MEC for Education, Training, Arts and Culture of the Northern Cape Provincial Government and another [1999] 4 All SA 590 (NC).

Minister of Education v Doreen Harris [2001] 4 SA 1297 (CC) 11 BCLR 1157(CC).

8 Changing the hegemony of race in schooling

The task of decolonising

Fieldnotes and reflections: schooling in the town

It's early evening on one of my research visits to Carnarvon, and I'm sitting in a restaurant talking to two local white men, a farmer and a businessman, both of whom have children of school-going age. Outside, two small Coloured boys dash towards a parking car with hands outstretched to beg. 'You see', says one of the men, 'I won't have my children walking around in the street like that. I won't send them to school with those children'. The second man gestures his agreement. I ask about alternatives and am told that their children are being home-schooled, at boarding school, or at the local support centre (which caters to home-schooled students). The quality of instruction at the local support centre is not a major concern for these young men, at least not at the primary level. They are confident that their children will be fine when they go away to boarding school for secondary education. Clearly, other things are more important to them than the formal curriculum.

I visit the support centre for white children who are registered for home-schooling, which is in an unmarked house in a suburban part of the town. Inside, small groups of children sit quietly at desks in the different rooms, and some are sitting in a group in the garden. I realise that I have seen no white children in the streets or public places in the town during my research visits.

Visiting the support centre, I'm told that 'Here, they stay "children" longer. They are not "mall children"'. I ask, 'What do you regard as most important in education?' Overwhelmingly, the answers relate to language and culture: to speak properly, have self-confidence, have moral values, religion, discipline, and routine. It's about 'the way you do things and the way you think'. Yes, there are disadvantages to attending a small support centre, most notably that extramural activities suffer, but 'you learn to create your own entertainment'. Many of the town's older children go to boarding school, I'm told, but 'boarding school doesn't suit every child'.

In discussion with white teachers at the town's two schools, I'm told that the teachers themselves prefer the support centre for their primary-age children.

One admitted that there were mistakes in the teaching of mathematics, but she would not consider the alternative of sending her child to the school where she taught. 'She is 'too gentle /*te saggies*', says the teacher; 'She and her brother will go to boarding school when they're old enough'. I'm told by a number of interviewees in the town that boarding school is a longstanding tradition for Karoo farmers. There are many excellent alternatives to choose from among the old, prestigious Afrikaans-medium public schools in the south-western Cape. These are schools that have 'culture', 'strong discipline', and 'sport', a farmer tells me. One of the teachers referred to a hierarchy in status (and no doubt cost), with boarding schools in places such as Oudtshoorn and Wellington being more accessible than the prestigious 'great schools' of Paarl and Stellenbosch. Family tradition will play a part in the choice of schools, as will the child's own interests, and the quality of care offered by the school's hostel and 'hostel parents / *koshuisouers*'. As one of the white parents pointed out with regard to boarding schools, 'The main difference is that whereas before they were half an hour away, now they're further away'. I'm told that young people 'spread their wings', and indeed 'wings' is a metaphor of mobility used by several townspeople I speak to. The problem is, though, that those with wings will probably not come back to Carnarvon, except to visit.

'It's not just the threat of race', I'm told by the pastor of the Uniting Reformed Church, 'but really the threat of poverty'. In his view, what the white townspeople illustrate is how poverty is perceived from the perspective of bourgeois culture. 'It's a class struggle, and race is assimilated'. In their treatment of others, these white townspeople have a '*laager* mentality' where they assume they are 'the *baas*'.[1] Several other Afrikaans-speaking interviewees mention that Afrikaners have '*stywe nekke*' (literally 'stiff necks'), referring to an unbending attitude when their control is challenged.

The primary school teachers appear to have easy collegial relationships at work, but there is almost no contact across designated race groups outside of school, though teachers often live in the same part of town. Several primary school teachers mention an incident where a white teacher held separate gatherings for a birthday celebration, one for Coloured colleagues and a larger one for white friends. This was called out as racist by a young Coloured teacher who expressed her hurt and disappointment. 'But', said one of the white woman teachers who attended the main gathering, 'they probably wanted to protect the teachers from their [white] friends around the *braai* [2] – you know how things can be'. It strikes me that this is probably one of the first times that a person of colour has challenged a colleague's behaviour as racist – partly because 'niceness' is one of the norms that these Afrikaans-identifying white middle class Christian women uphold. Later, another white teacher tells of the same pattern of segregation in a farewell party held for her by white colleagues – a deeply discomforting experience. In her thank-you speech, she took the opportunity to affirm her choice to embrace the Coloured community and was met with tears from her colleagues. Tearfulness as a response to confronting issues on race is a recurring response.

Segregation is perhaps most striking in the two major reform-tradition churches in the town, the *NGK* and its former *Sendingkerk* (mission church) now the Uniting Reformed Church, each with its old, classic buildings around the corner from each other. There is no doubt that these segregated churches are a major sign of the social dynamics of the town, since both have solid and committed congregations. I learn that some of the white teachers and their friends no longer formally attend the *NGK*, which is identified with a certain closed-mindedness and with older apartheid-supporting congregation members. Instead, they turn to home churches as an alternative, and some support evangelists such as Angus Buchan. I note the similarity between this response and the decision taken by white teachers to not send their own children to the schools they themselves teach in – withdrawal rather than engagement.

As I reflect on the comments about 'mall children' and the young boys who were seen begging outside the restaurant, it comes to my mind that there are over 1,200 children in the primary school and nearly 450 young people in the secondary school. Given these numbers, those begging on the streets (the 'mall') are a tiny handful of the town's children and young people. From my observations of children and young people walking to both schools at the start of the day, it's clear that poverty is a companion, but not all are poor. Nor do all poor children hang around in the streets in the afternoons and evenings. All of the town's children speak Afrikaans as home language. And the two major churches, racially segregated as they are, are based on the same Christian Reformed tradition. It seems to me that differences are highlighted by members of the white community, while similarities are downplayed. The Coloured community seems to be in no position to challenge this.

Researching race and class in Carnarvon

In interviews and informal discussions in the town, it was not easy to uncover the details of what had happened over the twenty years since education policies were changed by the post-apartheid government. In fact, it took three visits to the primary school on separate research occasions before the principal told his staff that they could trust me with their stories. And though a number of teachers told me of their experiences, I was aware of merely touching the surface of what had taken place. I soon learned that asking questions about the school touched a raw nerve for many of the people I approached. As well as avoidance and tears, I was met with overt resistance. A former chair of the governing body agreed on the phone to meet me on a street corner at 6 pm one evening, saying that he would speak to me, but 'first, I must look you in the eye'. (I did, unfortunately, go to the wrong street corner, but later met up with him outside the *NGK*, which was holding evening services for Pentecost that week – but that's another story.) One of the business-people, a prominent member of the church community, commented each time I returned 'Are you back again?', asking on one occasion 'What is it that you want to know about us?' Indeed, only when accompanied on two fieldwork visits by a trusted former teacher in the town would some of the people – white and Coloured – talk about certain events. Over time, overt suspicion gave way to a kinder acknowledgement among a number of the townspeople I spoke to, but people's reserves seldom dropped. As I put together a narrative of events around school restructuring, I became acutely aware that many lives had been touched by the events, not always in ways that they were prepared to talk about. In fact, one of the former deputy principals of the Coloured school said bluntly, 'I am happy to talk to you about anything in the town, but not the school'.

During the research process, I struggled with the complex, emotionally charged context I had unknowingly selected in deciding to do fieldwork in this town. I needed to accept that it is not possible for a researcher to sidestep the poverty, racism, stigmatisation, and cross-cutting emotions that are evident in a South African town like Carnarvon – however they are explained. There are no neutral spaces in research, and its power dynamics cannot be ignored. Exploitation hovers uncomfortably close when some people are able to enter the lives of others and use what they find to their own purposes – as feminist researchers pointed out many years ago. An outside researcher has the freedom to write and move on, to leave any disruption they may cause behind them without being required to mediate the consequences. Researchers are in a privileged position when they sit as observers in other people's classrooms, or interview individuals about their views while promising confidentiality and anonymity. The ethics of reportage is not a simple matter, particularly when a place is as identifiable as this town is and cannot be anonymised.

Yet there are also bigger issues of social justice and scholarship to consider. On reflection, I concluded that residents in the town know far more thoroughly than any outside researcher about the lived experiences of discrimination and my research account should not mute this. For me, the research task would be to engage in ethical and scholarly ways with the field – to engage with intellectual rigour, with concern for the fragile civility of social arrangements, and with an ethics of care. The challenge would be to write about social relations of discrimination and abjection in ways that do more than simply reiterate, amplify, and perpetuate them. It would be important for me to avoid being an echo chamber for the prejudices of others or a self-righteous judge of their lives; and it would also be important to present an analysis for scholarly debate.

Initially, I had been frustrated by the reserve I encountered, by people's reluctance to speak and their obvious avoidance of topics, by the silences I seemed unable to penetrate, by gaps in the accounts I was able to access. Over time, as I reflected more about the power dynamics of my research, I reconceptualised my task as a 'contact zone' activity (as described in Chapter 1). I realised that if I shifted my understanding of the relationship between myself as researcher and 'everything I encountered' to acknowledge these power dynamics, a different picture could emerge. The discomfort I experienced in the 'push back' of people against my questioning was indicative of power relations that I needed to respect if I was to understand my context. Focusing on the goals of my research, I realised I could learn a lot without asking, and indeed I found I had gathered more than enough for the analysis I was crafting to meet my research aims. In the discomfort of this experience of research-as-contact zone, I came to listen differently to what I heard, and to view differently what I saw.

Since racial terminology saturates the material of this chapter, it is valuable to pause and reflect on the way 'race' continues to signal power relationships in South Africa, even though it is widely recognised as a social construction. Deborah Posel's (2001) work on apartheid and race provides useful insights, showing how apartheid's mode of rule involved the systematic bureaucratisation and normalisation of race. Apartheid's architects, she suggests, 'recognised explicitly that racial categories were constructs, rather than descriptions of real essences', but this did not detract from their powers of classification:

> [Even if they were] constructs, these categories were powerfully rooted in the materiality of everyday life. The ubiquity of the state's racial designations, and the extent to which they meshed with lived hierarchies of class and status, meant that apartheid's racial grid was strongly imprinted in the subjective experience of race. This is not to say that those subject to processes of racial classification wholly acquiesced to them, nor that these categories were fully or simply internalized ... Yet it would be difficult to deny the extent to which the demarcation of South African society into whites, Indians, coloureds, and Africans has been normalized - for many, a 'fact' of life.
>
> (p. 109)

The practices of racial classification continue to have salience in South Africa, whether manifesting as the normalisation of race (and its accompanying attributions of superiority and inferiority), or the efforts to build non-racialism (as seen in the post-apartheid constitutional arrangements). With regard to the town of Carnarvon and its schools, the normalisation of race appears to be hegemonic within the white community, and there is little evidence of change towards non-racialism.

In exploring the notions of race so evident while researching this town, Charles Taylor's (2004) notion of the social imaginary provides a helpful framework.[3] Taylor points to the day-to-day assumptions people have about social arrangements as distinct from their theoretical or intellectual reasoning. In his words:

> I am thinking ... of the ways in which people imagine their social existence, how they fit together with others, how things go on between them and their fellows, the expectations that are normally met, and the deeper normative notions and images that underlie these expectations. (p. 23)

For Taylor, the social imaginary refers to a breadth and depth of shared, implicit understanding of the way the world operates – the background understanding that informs what people expect of each other, of 'where we stand in space and time', of 'where we stand in our history, in the narrative of our becoming', of where we stand in relation to other nations—in short, 'our whole predicament in time and space, among others and in history' (pp. 27–28). Taylor's work is grounded specifically in a study of Western modernity, but his general concept has value in describing the views expressed in this research study of a small and marginal town in South Africa. What needs to be foregrounded in this case, though, are the power relations which hold this particular imaginary in place, even when it is not endorsed by a large proportion of the community.

In this chapter, I explore the social imaginary of the town with regard to its schools – the taken for granted assumptions about everyday social arrangements, and how these play out in schooling. The chapter describes the particular images of difference that emerged in my research, showing how these framed the dispute over control of the schools described in the previous chapter. The dispute over the schools, I suggest, had the effect of heightening divisions in the town rather than bringing people together on a common project, which had been part of the policy vision of 1994. This signals the limits to change through policy, particularly when people have the agency and resources to resist the intentions of policy.

The previous chapter showed how the ambitious and ambiguous system of school governance – a product of the power-plays of constitutional negotiations – left significant political interests in place so that individual schools have been able to resist the equalising intentions of government policies.[4] This chapter shows the notions of race and difference that saturated discourses about the schools, keeping alive the inequalities of the past. And it shows the consequences of the withdrawal of white resources from the schools, leaving them as examples of the bare provisioning of the post-apartheid dispensation, where schools rely on state funding alone. Challenging this particular social imaginary of difference is crucial if the decolonisation of schooling in South Africa is to be more than an impossible dream.

Difference and the social imaginary of the town

The fieldnotes and reflections at the start of this chapter suggest a distinctive set of taken-for-granted assumptions about social arrangements, particularly among the white townspeople interviewed for this study. Their comments reflect a particular social imaginary, whereby the experiences of life in the town are assembled into a set of meanings that are assumed to be normative. What is strongly evident is a trope or figurative depiction of social difference. In this trope, racial classification has unquestioned status as a central marker of difference – it is a 'fact of life'. Race is easily linked to cultural difference; race-culture is seamlessly linked to poverty; and both are associated with undesirable behaviour. This set of views then works recursively as justification for taken for granted assumptions about social distinctiveness and accompanying notions of superiority and inferiority. Where behaviour that is considered undesirable is linked to racial difference, and this in turn is linked to cultural difference and poverty, the trope becomes self-perpetuating. In the sections that follow, I expand on these points, drawing on a range of formal interviews and informal conversations with townspeople.

In my formal interviews and informal discussions with white townspeople, common sense assumptions about differences of race-class-culture-behaviour were widespread as a connecting trope. Almost to a person (but with few notable exceptions), white interviewees assumed the obviousness of social difference that justified social separation, and these assumptions seemed resilient to challenge. When self-understandings of superiority (that had no actual basis) were made explicit in the course of discussion, or the evidence for assumptions was questioned, interviewees conveyed no need or desire to change their expressed views. A number of people humorously shrugged these off as an aspect of Afrikaner identity (such as '*stywenek*'); others showed disinterest, changed the topic, or simply fell silent. Indeed, my noticing such things as an interviewer was treated as an indicator of my outsider bias – another means of distancing from the subject. In the case of the school in particular, the boundary of difference seemed to be actively maintained with no interest at all expressed in dissolving it.

As a reflection of the power relations of the town, connotations of superiority/inferiority infused the trope of difference for white interviewees. 'We' (the white community) set the standards of what is acceptable to 'us', and it is important to be 'in control' of social arrangements in the town. Cultural practices are taken to signify major differences (for example how 'we' conduct our funerals compared with what 'they' do) as well as minor ones (how 'we' entertain ourselves with 'our' friends at home compared with what 'they' do at picnics). With regard to schooling, difference is expressed in terms of bodily comportment, appearance, gender-specific norms of behaviour, 'manners', and so on. Externalities of behaviour are then viewed as expressions of moral worth, again assumed to divide 'us' from 'them'. The soft edge of this trope of difference is expressed in the need to look after one's 'gentle' children, or to protect one's Coloured colleagues from the racism of one's friends at social events. The hard edge of the trope is evident in the willingness to allow other people's children to attend the overcrowded two-schools-on-one-site in the Coloured township, while one's own children attend a half-full school across town, with smaller classes and replete with cultural opportunities and sporting facilities.

In this particular social imaginary, white interviewees glossed over the history of apartheid and its structural inequalities. So normalised was racial difference, that the spatial segregation of the town was also normalised, as were the links to poverty ('them') and ownership of farms and enterprises in the town ('us'). A number of white interviewees emphasised that 'politics' had intruded into the town and its schools with the changes of 1994, not acknowledging the politics of the past. New local government arrangements had replaced the town's functioning institutions with a cumbersome, ineffective, and sometimes corrupt and self-serving bureaucracy. A former mayor complained that people who 'had no idea of how to run their own accounts were now in charge of the town's budget' – but shrugged off my suggestion that apartheid might have had a role in new appointees' lack of experience. In this social imaginary of white competence and even-handedness, the loss of control over the school had threatened its quality, and the church needed to be kept apart from politics. In this normalising imaginary, the contestations of history are rubbed over to leave a clear and unencumbered picture of 'natural' white exclusivity, while the structure of ownership of resources in the town and the historical inequalities cemented by apartheid are not questioned. The institutions of the town – its municipality, school, church – are not there for equal sharing. 'Politics' was regarded by these interviewees as a challenge to white control, while apartheid, as their baseline, was not viewed as political at all. No one specifically named 'apartheid', referring rather to 'the past'.

Obviously, the desirability of this social imaginary is not shared by everyone in the town, particularly since it works as a justification for maintaining inequalities. Not surprisingly, members of the Coloured community articulated different experiences of its exclusions. In interviews, Coloured teachers did not hesitate to name their treatment as 'stigma', and they expressed concerns about Coloured children being cast as 'inferior' in the town. Many of the experienced teachers had lived or worked in other places as well as Carnarvon, and they drew comparisons with what they knew of towns and schools elsewhere. They spoke of the 'stubbornness' of the white community in this particular place; of its 'lack of generosity'; and of its obvious 'anger and resentment' about changes to the school and municipality. In their words, white townspeople are 'exceptionally racist'; the Afrikaner farming community and townspeople 'don't want to break down barriers', but instead form a 'tight community' who 'protect their own'. A number of teachers made a point of saying that they regarded their qualifications and classroom capabilities as being equal to those of their white counterparts and they lamented the absence of white students in the town's schools. A number of them had sent their children to the former Model C school in the past, paying fees for the benefit of smaller class sizes and extramural and sporting opportunities not available to the overcrowded and lesser-funded

Coloured schools sharing one site. They nonetheless expressed their satisfaction with the restructured arrangements, in spite of the loss of the schools' middle-class advantages – a point I return to later.

As educated people with stable incomes, Coloured teachers had a measure of mobility and were able to live in the town's suburbs when the Group Areas Act was lifted. Not so the bulk of the community that their schools served, most of whom lived in and around Bonteheuwel, locked in by apartheid's remaining spatial geography and their lack of resources. Teachers spoke of limited work opportunities beyond the service sector, which itself was relatively small, mentioning also that the abattoir, supermarkets, and petrol stations were the main employers of young school leavers. They spoke of limited social opportunities for young people; of the 'rough behaviour' of boys in particular; of youth pregnancies ('it's not only girls', I'm told; 'boys need to show their peers that they have fathered a child'); of the evidence of foetal alcohol syndrome among the children at school. Teachers spoke of the difficulties of involving parents in their children's education even though these young parents had been to school themselves, often in the town. They spoke of having few resources for extra-mural activities and, as a result, limited participation in sporting and other competitions with schools elsewhere – in contrast to the opportunities that had been available in the former white school. In these circumstances, sponsored competitions offered important outlets, without which students would have few occasions to leave town. For these teachers, the trope of difference was experienced as a sense of limited opportunity, of power and resources held tightly by the white community and out of the reach of the Coloured community, and of the narrow horizons of possibility for many of their students (and possibly themselves, their colleagues, and their own children). As one of the teachers perceptively noted, 'aspiration requires a certain confidence', and the circumstances of the town did not nurture young Coloured peoples' aspirations for a more agentic future.

In short, poverty related to race has resulted in precarious lives for most people, with limited alternatives in the geography of this place. Unemployment translates into living on social grants, with stigma and lack of opportunity feeding a sense of passivity. It is surely not surprising that sex, alcohol, and drugs are outlets for leisure/pleasure for young (and not so young) residents in the town. Several teachers pointed to the obvious fact that there are more liquor outlets in the town than sport and recreational facilities. The principal of the *Hoërskool* keeps track of students who have bursaries to study further (mostly from the SKA). He tells me that some young people form 'Carnarvon communities' in Cape Town and elsewhere (mentioning the Cape Town Coloured communities of Atlantis and Elsies Rivier as examples of places). Many return after a year or two away, particularly if they cannot find work.

Considering the social imaginary of this place, what is striking is how little is shared in common. There is almost no sense of 'ours' in the town, of common endeavours to build social connection beyond the bounded racialised communities with their obvious power differentials. In the picture that Coloured teachers sketched of life chances for young people in the town, white people were notably absent. It seems that the apartheid past has saturated the present and clouded the possibilities for imagining a different future.

Constrained by the limited opportunities of place, stigma operates in recursive ways. People are positioned in a social hierarchy based on historical patterns of possession and dispossession, and those who are privileged turn to despise those who are not, blaming them for being where they are in the social hierarchy. This move is one of abjection. Moral status is attributed to social privilege, and disadvantage is viewed as personal or cultural failing.

Contesting the schools

Against this sketch of the trope of social difference in the town, the struggle over the school exemplifies how difficult it may become to share spaces when difference hardens into contestation and brings strong emotions that are difficult to resolve. In expanding on this point, it is worth noting again how difficult we found it, as outside researchers, to open discussions with townspeople about the schools. We came upon a great reluctance to speak about the events that had led to change in the schooling arrangements and why the white community would not send its children to the schools. One of the white community members summed up what we experienced, saying: 'When you ask people about the school, they either burst into tears or keep quiet'. We found this reluctance particularly marked at an informal gathering for tea that we hosted with a group of Coloured and white teachers from the primary school, some of whom had moved across from the former white school in 2007. While individual teachers had been willing to talk to us informally in one-on-one conversations about their experiences of the change, no one was prepared to speak about their experiences in front of the others in the group. The atmosphere was collegial and friendly, but the unwillingness to speak openly was pervasive. Clearly, sensitivities around these events and their racial inflections run deep.

This study was concluded more than ten years after the schools were restructured in 2007, as described in Chapter 7. In that year, a group of Coloured parents had organised under the banner of '*Besorgde Ouers*' ('Concerned Parents') to contest the SGB elections of *Hoërskool* Carnarvon and had won control, much to the shock of the white parent body. This had cleared the way for the schools in the town to be amalgamated, ending a ten year stand-off between the white-controlled SGB and the Northern Cape Education Department. In the process of amalgamation and restructuring, *Hoërskool* Carnarvon was closed; pictures and memorabilia were removed, except for the honours board in the school hall; and a new uniform and school badge were adopted for the secondary school. (We learned later that the photographs and memorabilia had been gathered and stored in a room at the *NGK*.) The premises of *Hoërskool* Carnarvon were handed over to a newly constituted secondary school, while the Bonteheuwel premises (which had been custom built for a secondary school), were allocated to the primary school. The changeover happened over a weekend in April 2007, with staff and students transferred between the two school sites for the start of the next school week. Primary school teachers who had previously taught at *Hoërskool* Carnarvon spoke emotionally of moving their posters and learning materials across to a different classroom in the township school, where class sizes were larger and they did not know all of the children.

For both schools, this was an unsettling time, with staff positions realigned and class sizes changed. Despite the passage of time, feelings still run high when the issue of the school is raised. It took six research visits to pull together the narrative of events put forward here from the perspective of different participants. In what follows, I have selected two interviews as illustrative. The first is an interview with one of the *Besorgde Ouers* who had been involved in planning and carrying out the changes at *Hoërskool* Carnarvon. Extracts from the recorded interview are presented. The second is an interview with a white farming couple whose children had attended *Hoërskool* Carnarvon at the time, and who had played an active role in blocking the restructuring. Extracts from notes taken during the interview are presented.

- **An interview with one of the 'Besorgde Ouers':**

  Everything began with the textbook levy [*boekegeld*] that the school introduced in addition to fees. We thought that fees should include books. That's where the story began.

  We appealed to De Aar [the District Office] to intervene. But *Hoërskool* Carnarvon was a Model C school with structures that couldn't easily be shifted. We complained to the Province. And the *Hoërskool's* resistance intensified our resolve. With the new SGB elections we lobbied, we held meetings with parents in the SASSA Hall. We contested elections and won positions on the SGB. They [white parents] were shocked at the elections. We had manifestos, and they didn't anticipate the numbers who would attend. With the *Besorgde Ouers* on the SGB and the muscle of the Department, amalgamation went ahead.

  The [secondary] school in Bonteheuwel was built for 800 students, and we had 1800 [when the primary school was moved onto the same site]. We wanted better learning conditions for all children in the town, not just a select few. It was a great relief that children could have education under better conditions. We had had friction [on the single site] and were at odds. The primary school had felt under pressure because the administration block was held by the secondary school, and they had to operate from classrooms.

When Carel van Zyl [the primary school] burnt down, we had no idea we would be on the premises of *Sekondêr* [in Bonteheuwel] for 17 years.

We simply wanted equal opportunities for all children. Whites have a democratic right to send their kids to the support centre.

I am completely happy with how things transpired. Things are relative. For me, I can only see how much better things are for the primary school. I think for example of the toilets!

We knew that *Hoërskool* Carnarvon always won court cases so it would have been a lost cause to take them on without the Department. So we saw the textbook levy as a point to take up. It was a small matter, and no-one would give in. If they had just given in, there wouldn't have been a follow up. For the *Besorgde Ouers* we had a point to rally around.

- **Notes from a meeting with white parents:**

  On the coloured school in Bonteheuwel:

  Whites in the town said they wouldn't take their children over the river to the township school.

  The township was 'a no go zone'. It was unknown to us.

  We were frightened about what would happen to the children if they went there.

  The standards at the school were not the same. And the sports facilities were here [at *Hoërskool* Carnarvon].

  On the election night:

  That night changed everything.

  It was overwhelming.

  There was a mass of brown parents.

  On fear of 'the other':

  The communities were unknown to each other. We were strangers.

  The political system had changed, but ordinary people didn't know each other.

  At that stage, Coloureds were an unknown community.

  We were fearful of what would happen.

  We felt threatened.

  It was too drastic, too quick.

  People fear the unknown.

  On the feelings of loss:

  The loss of the school felt like an emotional blow.

  It was something [precious] that was our own. (*'ietsie wat ons eie is'*).

  You felt threatened because you were not in control. The moment that control was taken away, you didn't have trust in the system.

> How to accept change?
> The white community would do anything to help the school in the past. They won't support the school now.

Do you think *any* children should endure the conditions of Bonteheuwel's schools [with two schools on one site]?

> I can't answer that.

Hardening of difference

There is no doubt that the fight over the school hardened the dynamic of possession and dispossession that was part of the history of the town. In their determination to keep control of the school on their own terms, the white SGB played a zero-sum game and lost. The *Besorgde Ouers* were able to exercise power that their white counterparts had not anticipated they could possess in relation to schooling – the power to elect a new SGB which gave them equal rights. Interviewees on both sides of the dispute spoke of the 'hatred' that the other side felt towards them. White townspeople were not prepared to assist the school in any way and withdrew all the resources they legally could. Coloured teachers resented this withdrawal, mentioning that the Model C school had sold its bus and trailers, and removed a trust fund they had set up. They also mentioned the meanness of white business owners who benefited from the schools' presence (for example, the primary school feeding scheme) but would not support the schools with anything more than token donations. And they noted that the *NGK* was not prepared to direct any donations to the schools, though they had done this for the *Hoërskool* in the past.

Nonetheless, the *Besorgde Ouers* were satisfied with the outcome overall. In interviews, a number of them spoke of the excitement and even exhilaration of this moment where they had achieved equal status for all school children. Though Coloured parents who could afford fees had lost the opportunities offered by the former Model C school, the conditions for Coloured children in general had improved from the cramped and overcrowded two-schools-on-one-site, and this they regarded as a beneficial outcome.

For the white community, however, the sense of anger was almost inexpressible, as the following statement by a leading townsperson suggests:

> My children went to school here. But not my grandchildren. I cannot send them to the school here. It is not the race. It is something else.

The sense of dispossession on the part of white community members was confirmed by everyone we spoke to in the study, including those who did not share the feelings. The associated emotions were well articulated for us by a senior member of the *NGK*:

> The [white] people in the congregation have harboured ten years of anger about the school, not because they had to open the school to include

> others but about the way it was taken from them. They had no problems to open the school but everything including their children were taken from them. It is anger, it is not a political issue, it is an emotional issue. The town will not heal or reconcile until the school is healed. The high school [the site of the former white school] is the biggest problem. They are so politicised. The SGB members are all ANC. The whole town is politicised ... The whites do not want community actions that have anything to do with the school.

What is intriguing in this statement is the claim that 'their children were taken from them'. In fact, what was taken away from white parents was not their children but their power to control the school. The decision about how to respond to this loss of power was their own, the result of their own agency: to withdraw their children and instead use the resources they had to send them to a support centre or boarding school. Yet the sense of loss – and anger at loss – was palpable and extended beyond the school and into the town more broadly.

At the same time, it is worth noting that some of the members of the Coloured community in the town also expressed a sense of loss and anger at loss – in their case, loss of the land they had been given legal rights to in the past and could not retain because of poverty. The oral testimony recorded in the Uniting Reformed Church's *Gedenkblad* shows clearly that older members of the Coloured community remember their parents' earlier possession of farms now in white hands, and the names of the farms are listed in the *Gedenkblad* – names that appear also in Auret's land survey report of 1847. Community members also recalled their battle to retain their houses in *Die Bult* in the 1970s, when the Municipal Council invoked the Group Areas Act in an effort to have the houses demolished to clear the way for a 'white area' – a battle which the Coloured community won, with some white support. They also remember that during the repression of political activity in the 1980s, their pastor, Arné Leuvennink, had been detained without trial in the local prison, alongside the principal of the primary school, Benjamin Hoorn, pastor in another church.

In understanding how events may take on symbolic and emotional significance beyond their actual form, Andries du Toit's (2013) work on land reform has important resonances with this event in schooling. Du Toit, albeit writing in another context, talks of how emotions associated with larger and past events may become entangled in a particular moment in symbolic ways, and he outlines what happens when 'real-world problems get entangled with an essentially symbolic drama':

> Anger at the injustices of the past; fear of retribution; sorrow, fear and guilt about injustices caused; confused desires for redemption and vindication – all these were (and still are!) richly present for all participants. This means that the politics of South African identity formation are what Freudians would call deeply cathected: every event or act is pregnant with meanings infused by histories well beyond the intentions

> of the actors; every fact (however nuanced and complex reality might be) is available for construction as evidence for powerful and often reductive narratives of betrayal, retribution, bad faith, triumph, failure and so on.
>
> (p. 8)

In this case, the social imaginary of difference became hardened into strong emotions of anger and loss when cathected through the event of school restructuring. In a historical context of possession and dispossession, a sense of difference had moved easily into feelings of aversion, and aversion became hatred. Once cathected in a complex of associations, it seems almost impossible to resolve the issues as real-world problems.

The face of poverty: schools after 2007

Given the historical concentration of resources in white hands, the withdrawal of white parents from the town's schools left them depending solely on state funding for their operation. While operating in functioning premises with qualified staff, neither school had resources to provide much by way of sporting or cultural activities. Teachers and students shared the same language (Afrikaans), and in this regard Carnarvon's schools are better off than many in the country, since Afrikaans is one of only two languages that the curriculum supports for all school years. Yet the schools' results match the profile of disadvantaged schools in the bimodal performance of the system. For me as researcher, what was striking was the primary children's anticipation of school lunches provided by the national school feeding scheme (though some brought their own). Also painfully evident was the love and care, or absence of this, as teachers served this food to the children in their classes.

Carnarvon's schools: curriculum-in-place

The results achieved by Carnarvon's schools are set out in the accompanying Tables 8.1–8.3.

Table 8.1 Overall percentage pass rates, Pixley ka Seme District

Grade	*2103*	*2014*	*2015*
G 3	58.2	62.7	64.2
G 6	37.9	33.6	40.8
G 9	40.2	13.7	20.5
G 10	34.9	24.5	37.5
G 11	48.4	40.1	49.0
G 12	60.3	63.4	66.5

Source: Department of Basic Education.

Table 8.2 Results from Carnarvon schools, quarter 4, 2017[5]

Phase	*Grade*	*Total learners*	*% learners promoted*	*Grade average (%)*
Foundation (R–3)	R	94	100.00	69.57
	1	217	70.97	66.67
	2	162	95.68	68.61
	3	147	87.07	54.46
Intermediate (4–6)	4	209	72.25	47.57
	5	143	90.91	52.88
	6	105	80.95	44.47
Senior (7–9)	7	95	53.68	47.97
	8	157*	58.60	34.57
	9	80	83.75	43.34
FET (10–12)	10	71	67.61	38.55
	11	67	83.58	43.42

*Intake includes other primary schools in the district
Source: Information from the schools.

Table 8.3 Average grades for Senior Certificate, school and national for Mathematical Literacy, Mathematics and Physics

	Mathematical Literacy average%		*Mathematics average%*		*Physics average%*	
	Carnarvon school	*National*	*Carnarvon school*	*National*	*Carnarvon school*	*National*
2011	48	47	26	32	28	36
2012	48	46	26	35	28	38
2013	42	46	28	37	26	40
2014	50	45	18	35	26	38
2015	44	40	35	33	42	37
2016	44	40	36	34	33	39

Source: Department of Basic Education.

Looking across these results, what immediately stands out are the low percentage of pass rates in the district, the low and declining grade averages in the schools, and the low achievement levels in Senior Certificate. Although Carnarvon's pass rates at grade level may appear higher than the district's pass rates, its grade averages suggest that performance is similar. Promotion percentages dip at the curriculum transition points of Grades 1, 4, and 7 in the primary school. In the secondary school numbers drop from Grade 9 and percentage promoted increases, even though grade average continues to fall.

During my research visits, most primary teachers voiced their problems with the current CAPS curriculum.[6] They expressed frustration that the inflexibility of CAPS left them almost no discretionary space in terms of content coverage and pacing. As well as the heavy content demands, the fast pacing of CAPS and its insistence on assessment meant that they were required to keep moving forward in the curriculum, whether or not students had grasped what was being taught. The cumulative effects of inadequate and/or incomplete curriculum coverage were passed on from grade to grade, making the task of teaching the curriculum increasingly difficult. Secondary teachers, in turn, complained bitterly about ill-prepared students coming from primary schools in Carnarvon and nearby towns.

In discussions, the primary school principal and teachers pointed out a particularly concerning pattern caused by the national assessment policy on promotion, where students may fail only one grade per phase and thereafter are automatically promoted to the next grade. This is evident in the promotion percentages, where Grades 1, 4, and 7 – grades of phase transition – have dips in the percentages promoted as a result of the national requirements of failing once only per phase. In Grade 1, the majority of children encounter classroom learning conditions for the first time, and in Grade 4, the curriculum structure changes with the move from the Foundation Phase to the Intermediate Phase. Grades 4 and 7 typically show not only the lowest percentages of promotions, but also the lowest grade averages (not that any of the grades produces high scores). Grade 7 is another curriculum shift, to Senior Phase, and those who are automatically promoted into and out of Grade 7 are likely to fail when they move to the secondary school.

In effect, every class, including Grade 1, has a group of students who are not able to meet the curriculum requirements to pass that grade. In a positive scenario, if children fail Grade 1, repeat the year and pass (showing that they meet the curriculum requirements for promotion to Grade 2), they may well succeed in the following years. However, in another scenario, if students fail Grade 1, repeat the year and fail again, they will be promoted automatically until they reach Grade 4. It is likely that they will fail Grade 4, repeat and fail again, and then be promoted automatically to Grades 5, 6, and 7. If they fail Grade 7 twice, they will automatically move on to secondary school – without passing any grade at primary school. There they will stay until they reach the school leaving age of 16, or Grade 9.

I observed the effects of this promotion policy in operation in a number of different grade levels and classrooms. Class sizes at the primary school are large, and a common pattern I observed was for teachers to seat students in three rows or groups to enable more focused instruction for each group. Informally, I was able to observe students doing writing tasks in several classes, and differences in their work were striking to see. Perhaps most stark was a class that was repeating Grade 4, where all of the students struggled to form their letters in a writing task.

The combination of the promotion policy with a fast-paced and content-heavy curriculum that teachers struggled to cover means that classrooms are in effect 'multigrade'. In Eric Schollar's (2018) view, this 'multigrade effect' is the hidden variable of poor performance in schools across the country, where teachers must work with students who are in the same class but at different levels of achievement in relation to the curriculum requirements for the grade – some of them not achieving basic literacy and numeracy. Drawing on his research across different contexts, Schollar comments as follows:

> The inescapable conclusion is that the great majority of South African classes on a national scale have become, in effect, multi-grade in terms of ability and that teachers are faced with an enormous range of learner abilities in mathematics, science and other subjects, from the virtually innumerate or illiterate to the small pool of the genuinely competent in every class they teach. This makes it next to impossible to teach to, or assess, all of the learners at the appropriate grade-based assessment standards of the curriculum and the likely result, in practice, is that learners of any level of ability are not exposed in full to the conceptual progression made explicit in the national curriculum.
>
> (p. 118)

This situation is compounded by the fact that there is no provision for special needs education in schools like these. Students who have special learning needs are folded into mainstream classes, with no additional support. A difficult-to-mention issue for Karoo schools in particular is the high rate of foetal alcohol syndrome recorded in these communities, for which no schooling provision is made. In geographically dispersed areas such as the Northern Cape, district support is not 'thick' unless there are major problems. The district office for Pixley ka Seme is located in De Aar, 200 kilometres away from Carnarvon, and these schools are not the furthest away in this district. For both Carnarvon schools, each year of my research began with shortages of learning materials, a problem confirmed by a district official who spoke of difficulties with procurement and retrieval in schools such as these, as well as struggles with maintenance for which there is minimal financial support from the government.

Failure becomes the norm

Across the country, schools in conditions of poverty show similar patterns of failure, repetition, and dropout to those of the Carnarvon schools, albeit with different profiles. Teese and Polesel's (2004) distinction between 'fortified' and 'exposed' schools has relevance in illustrating this pattern. Where schools are supported by the material and cultural resources of their parent bodies, they stand fortified against the demands of the curriculum, while schools with bare resourcing in conditions of poverty stand exposed. Class sizes are larger, opportunities for supplemental teachers are few, and there are few resources for curriculum supplementation or cultural and sporting activities. Whatever the combination of reasons, the fact is that schools in communities too poor to pay fees generally perform substantially less well than their fee-paying counterparts. Estimates are that 'by the fifth grade the educational backlog experienced in historically black schools is already equivalent to well over two years worth of learning' (Taylor, 2011, p. 16).

In short, schools in poor circumstances – the majority of the country's schools – consistently perform poorly on all measures. They are required to accommodate failure as part of 'normality' and must absorb this as the way things are. Though the label of 'failure' is attached to individuals and schools, it is actually the schooling system that fails to provide the necessary conditions for success for the majority of its students. What is interesting in the case of Carnarvon is the clear picture of the consequences of the withdrawal of middle class resources from the town's schools, leaving them as 'bare schools' relying on state provisioning for their operation.

Reflections on change

As discussed in earlier chapters, the political changes at the end of apartheid did not significantly interrupt or end established patterns of economic ownership and geographic segregation, which remained in place. Indeed, the negotiated settlement shows that it is possible to change political arrangements and establish a common citizenship with equal rights for all, while leaving social and economic inequalities largely intact. And it is possible for a social imaginary of difference to continue long after formal equality is proclaimed. In the case of Carnarvon, racial patterns of ownership predated apartheid, and social arrangements in the present have not shifted the material resources and associated social power. Nor have they shifted the social imaginary of superiority/inferiority, notably among the white community.

In the case of the schools, issues of power and control were at stake in the restructuring process. The white community had been prepared to facilitate a limited sharing of the school's resources as long as they could set the terms and remain in charge. For ten years after the South African Schools Act was passed, the white SGB of *Hoërskool* Carnarvon was able to run the school on its own terms, in effect using the technicalities of the law against the spirit of the law. When this power was taken away from them, they were able to boycott the schools, establish alternative arrangements for themselves, and use mobility – the ability to send their children elsewhere – to their advantage, but also to their loss, as their emotional responses show. Given that the town has no post-school education provisions, mobility is key to educational futures. Having 'wings' makes a difference to schooling opportunities for individuals – but it is not without its costs to the broader community if mobility means leaving for elsewhere.

White parents' withdrawal of resources from the schools confirmed the stark realities of what post-apartheid schooling provision offered to their fellow townspeople, and their withdrawal suggests that they were not prepared to have this level of provision for themselves. Nor were they willing to use their resources – economic and social – to build an alternative that all in the town could share. What they saw as a consequence of minimal state resources were the limited forms of 'schools in poverty' – that bare provision in no-fee schools may mean forty students in a class; many of them neglected and hungry; without school-related cultural capital from their homes; with minimal opportunities for sporting and cultural activities; and with minimal resources to maintain premises. Though these conditions were unacceptable to the white community, the restructured arrangements in fact brought an improvement for the two Coloured schools that had been cramped onto the same site, and the *Besorgde Ouers* were satisfied with the equality they had achieved.

In interviews, the white community seemed overwhelmed by emotions of anger and loss, with little concern that their own withdrawal of resources from the schools had compounded the situation and contributed to schooling conditions that they were not prepared to have for their own children. Narratives of past practices could justify their actions – 'we've always sent our children away', and 'we've always taken particular care to nurture our children'. While these narratives of the past may be true (as previous chapters have shown), it is also the case that they have become the template for the present, with no sign of the larger questions of future schooling for the good of all in the town being addressed. The town as a whole suffers the loss – themselves included.

In the case of schools in Carnarvon, the unwillingness of those with resources to share with others reflects not only the will to control, but also the importance of social class in the imaginary of difference – what children in poverty bring with them to school. The proximity of poverty brings not only abjection, but also the fear of their own children being pulled into its effects. It is easier, in these circumstances, to maintain a trope of difference than to venture into the unknown of a shared future. It is simpler for communities such as this to continue to use racial designation as an indicator of social standing than to confront the consequences of equal sharing of public resources, or, alternatively, contributing their own resources for the benefit of a common community.

What might be entailed in shifting this social imaginary of difference towards one oriented towards a shared future? Here, Jean-Paul Lederach's (2005) reflections on the 'moral imagination' are worth considering. Drawing on decades of work with societies moving from conflict to peace, Lederach poetically outlines the mind-shift required for major social change:

> Stated simply, the moral imagination requires the capacity to imagine ourselves in a web of relationships that includes our enemies; the ability to sustain a paradoxical curiosity that embraces complexity without reliance on dualistic polarity; the fundamental belief in and pursuit of the creative act; and the acceptance of the inherent risk of stepping into the mystery of the unknown that lies beyond the far too familiar landscape of violence.
>
> (n.p.)

What Lederach signals is a profound shift in how people regard each other. This involves letting go of fixed notions of 'us' and 'them' and a willingness to creatively explore who 'we' are in a shared existence. It is risky to let go of what feels familiar in patterns of daily life, to relinquish fixed identities as the basis for personal exchange, and to surrender the will to be in control and trust others. Of course, social change does not require the same adjustments of everyone involved; nor is it necessary for everyone to embrace the envisioned social order with enthusiasm – grudging compliance is all that is required. Yet without such shifts, change becomes almost impossible.

That said, a completely unanticipated shift in the historical patterns of possession and dispossession comes from an outside source: the arrival of the Square Kilometre Array into the vicinity of the town, to be discussed in the next chapter.

Notes

1 '*Laager*' refers to the fortress-like inward-facing encampments of ox wagons in trekking history – a common metaphor for Afrikaner closed-mindedness. '*Baas*' would translate as 'master' rather than 'boss' – a hangover from colonial labour forms, including slavery.
2 Direct translation would be 'barbeque', and in South Africa this is a particular outdoor ritual.
3 It would also be possible to explore this with Gramsci's notions of hegemony and common sense. In using Taylor's work, I am aware that his is a study of Western modernity, not intended to be universalized. Nonetheless, his framing of the imaginary is particularly clear and I use it for this reason.
4 Specifically, the South African Schools Act (1996), which gave significant powers to school governing bodies (SGBs), on the basis of which the SGB of the former white school in Carnarvon was legally able to resist restructuring for ten years (1997–2007).
5 A single year is presented for ease of analysis. While there are variations year by year, the overall patterns are relatively stable.
6 CAPS – Curriculum and Assessment Policy Statement – national curriculum.

References

Christie, P., & Monyokolo, M. (Eds.). (2018). *Learning about sustainable change in education in South Africa: The Jika iMfundo campaign 2015–2017* (pp. 99–124). Johannesburg: Saide.

du Toit, A. (2013). Acts, imagined landscapes: Reflections on the discourses of land reform in South Africa after 1994. *Journal of Agrarian Change, 13*(1), 16–22.

Lederach, J. P. (2005). *Moral imagination: The art and soul of building peace*. New York: Oxford University Press. Oxford Scholarship Online. Oxford University Press. doi:10.1093/0195174542.001.0001.

Posel, D. (2001). Race as common sense: Racial classification in twentieth-century South Africa. *African Studies Review, 44*(2), 87–113.

Schollar, E. (2018). Curriculum management, improving learner performance and the rise of multi-grade classes: A tangled web of challenges to the design, operation and evaluation of educational development programmes in South Africa. In

P. Christie & M. Monyokolo (Eds.), *Learning about sustainable change in education in South Africa: The Jika iMfundo campaign 2015–2017* (pp. 99–124). Johannesburg: Saide.
Taylor, C. (2004). *Modern Social Imaginaries.* Durham; London: Duke University Press.
Taylor, S. (2011). Uncovering indicators of effective school management in South Africa using the National School Effectiveness Study. Working Paper, University of Stellenbosch Department of Economics and Bureau for Economic Research.
Teese, R., & Polesel, J. (2003). *Undemocratic schooling: Equity and quality in mass secondary education in Australia.* Melbourne: Melbourne University Press.

9 The SKA comes to town

'Big science' and development

Place and scale

It is not possible to write about the town of Carnarvon without mentioning the presence of the Square Kilometre Array (SKA). Nonetheless, it can safely be said that the government's decision to join an international 'big science' collaboration to build the world's largest ever radio telescope – and to build South Africa's share in the vicinity of Carnarvon – had nothing to do with the interests of the town, its residents, and their schools. Given that the focus of this study is on place, schooling, and inequality, it is simply not possible to do justice to the SKA as a massive science project and how South Africa came to play a leading part in it. Nor is it possible to say with much certainty what the presence of the SKA will mean for the town of Carnarvon and its schools in the future – or even whether they will survive in recognisable form. Much as the SKA offers a tempting story about the triumphs of science and South Africa's bid for international prestige, the ambit of this study cannot be stretched to include much of this story. Instead, the main focus here remains on the local place. The account of the SKA provided here is purposely gleaned from secondary sources and the SKA's own publicity materials.[1] My intention is to highlight contrasts between the activities of the SKA as a global and national project and the lived experiences of Carnarvon as a local place, showing their different scales of practice and contextualising these differences in national development debates.

Earlier in this study, I referred to the notion of affordances as 'what the environment offers the animal' in an interrelationship of complementarity. The Karoo environment affords a home to those who live in the town of Carnarvon – a focus for activities that include schooling, making a living in the complex and unequal social relationships formed in this sparse environment, crafting a sense of identity and meaning, and a holding some sense of history.[2] Simultaneously, though, the Karoo also affords an environment particularly suited to large-scale radio astronomy – high atmospheric transparency, minimal radio frequency interference, seismic stability, and low population density. From these different points of interest, the affordances of Carnarvon are vastly different and not necessarily compatible.

Put differently, for those who lead their daily lives and attend schools in Carnarvon as their home-town, the rhythms of daily practice are very different from the rhythms of daily practice that are generated by the many different activities of the SKA. Though these different rhythms encounter each other in the place of Carnarvon, each has its own logic. The image of ripples caused by stones of similar sizes thrown into water and crossing each other in different trajectories comes to mind, with ripples affected by the encounter but carrying on rather than disappearing. However, this image of multiple ripples changes dramatically when the scales are different, one set of ripples being caused by a single stone and the other by a landslide of rocks. (And if a water image seems inappropriate for this dry place, we can remember that the whole of the Karoo was once under water!) The SKA is a massive project, operating on multiple scales – global, national, and local – and its engagements with the town of Carnarvon are a small part of its overall interests and activities. Commensurability will always be a problem in understanding the impact of the SKA on Carnarvon and its schools.

The Lefebvrean image of rhythms intersecting and colliding in social space implies continuing change, and this chapter offers no more than a snapshot that captures one moment in a changing social space. While maintaining a steady focus on the local place as vantage point, I read the SKA's activities as operating on different scales: global, national, and local. Viewing these as akin to 'waves in water', I attempt to give some sense of the different logics of practice as they encounter each other and frame this encounter in terms of national development debates.

The SKA on different scales

On a global scale, the SKA began as a series of informal gatherings of international astronomers in the early 1990s, interested in building a large-scale array of radio telescopes to investigate the far depths of the universe and explore its origins. An array would be necessary, rather than a single telescope, because of the magnitude of the task, with 'square kilometre' referring to the receiving area of the total array – about 3,000 antennae – rather than a geographical area. In 2000, as impetus gathered, the Square Kilometre Array Working Group was set up, and through a competitive bidding process, its project headquarters were housed at Jodrell Bank near Manchester. Much later, in 2011, the SKA Organisation was established as a not-for-profit company, involving about 100 organisations from 20 countries, to be steered by an Inter-Governmental Organisation Treaty Agreement. As a global mega-project in 'big science',[3] the SKA is the product of massive investments of money and expertise from multiple countries – and costing a conservative estimate of EUR 2 billion (Wild, 2012).[4] South Africa was a latecomer to the astronomy gatherings, and something of a dark horse – an unknown and unlikely winner.

On a national scale, South Africa's decision to participate in this enormously costly project and then to bid for a leading role in hosting the telescope array in 2003 was in part the result of complex policy jockeying within the new post-apartheid government. While the provision of basic services to the majority of the population – education, health, housing, job creation – was the main priority of the new government, there were other issues to consider as well. Among many other policy orientations developed at this time, a strong lobby argued the case for science to have a significant profile, not least in redirecting capacities that the apartheid state had built in military and atomic science and reintegrating South Africa into the world science community. This lobby argued that science and technology were crucial for South Africa's future development, and ambitiously linked this thrust for innovation to the provision of basic needs – thus straddling two major and potentially competing interests. An oft-quoted section from the first White Paper on Science and Technology of 1996 stated the case for maintaining 'a basic competence in 'flagship' sciences such as physics and astronomy', on the grounds that 'Not to offer them would be to take a negative view of our future – the view that we are a second class nation, chained forever to the treadmill of feeding and clothing ourselves' (1996, p. 16).

During this period, South Africa shifted towards global neoliberalism, adopting GEAR as its economic framework and aspiring to participate in the growing knowledge economy. Achieving prestige in basic science and innovation also blended well with President Thabo Mbeki's 'African Renaissance' vision: to restore Africa's centrality to world history, lift South Africa's international profile, and secure a leading role in Africa. This confluence of interests provided the conditions for the SKA-South Africa to burst onto the scene.

The vision of South Africa achieving symbolic status as an international leader in science – and at the same time providing for the basic needs of its people – was seen to justify large budgetary contributions to astronomy as an area where South Africa had geographic advantage and an existing knowledge base. In 1998, the Mbeki government contributed generously to building the South African Large Telescope (SALT) with international partners in Sutherland, and in 2003 it funded the Karoo Array Telescope (KAT-7) to be built near Carnarvon as a proof-of-concept telescope to support South Africa's bid to host the site of the SKA telescope array. When South Africa was shortlisted along with Australia in 2006, funding was expanded to build the 64 dish MeerKAT as a pathfinder to compete with the Australian Square Kilometre Array Pathfinder (ASKAP). The government backed the argument that the exceptionally high levels of funding for MeerKAT (from different national budgetary sources) would place South Africa at the leading edge of astronomy, even if the SKA bid did not succeed.[5] In 2012, after stiff competition, South Africa and Australia were named as co-hosts for the telescope arrays, with the more compact low frequency array awarded to Australia and the extensive mid-frequency array (the larger share) awarded to South Africa and the eight African countries that were partners in its bid.[6] The award of SKA Phase 1 commits South Africa and its partners to building 133 additional telescopes, incorporating the 64 dish MeerKAT and extending to several other African countries.

In outlining this brief narrative of the SKA in South Africa (SKA-SA), it needs to be said that the decisions taken to build capacity in astronomy and participate in an international 'big science' project were political in every way. They were in part the result of skilled policy negotiations at national and international levels, with support mustered by individuals with science backgrounds and political connections.[7] Political leverage was supplemented by strong advocacy from the astronomy and cosmology community who were a group of relatively small but powerful scientists, and many others as well (see Dubow, 2018; Whitelock, 2004). Though the international SKA Organisation funds the bulk of the activities, South Africa's own financial contribution is enormous.

Viewed in its own terms, the achievements of the SKA-SA are impressive. Regular reports to the South African parliament and the material on its website give the public a snapshot of some of the multiple and complex activities bound up in a project of this magnitude and scope, as well as providing evidence that the SKA-SA is on track to deliver on its targets in this highly competitive international venture. The significance of these achievements, which place South Africa at a leading edge in science, should not be underestimated.

That said, questions remain about the purported links between 'big science' and development in the South African context, particularly since material benefits are likely to be long term (if indeed they accrue at all) and South Africa's development needs are overwhelming and immediate. Working with competing priorities is a necessary task for government, and it is perhaps too soon to reach conclusive judgement. Taking a long view, Saul Dubow's overview article on '200 Years of Astronomy in South Africa…' poses a set of central questions to be considered in relation to the SKA, science and development in South Africa:

> The biggest set of questions are very much a product of post-apartheid promises and expectations: given the huge investment in government resources, is the SKA likely to meet the social and developmental promises that constitute a crucial element of its prospectus and so meet local community expectations as well as those of the international scientific community? Exciting as the prospects of the SKA undoubtedly are, there are troubling indications that it may not.
>
> (2018, pp. 24–25)[8]

Turning then to the local scale – the vantage point of this study – the contrasts are sharp. At this level, closest to the actual MeerKAT telescopes, the importance of 'big science' and the links between science and development take on a different significance, with longer-term national priorities receding in importance as local concerns loom larger. From the local perspective, as the following section will argue, the SKA may be viewed as another wave in the sequence of the possession and dispossession of the land of its livelihood, established from the early days of recorded settlement. This then raises further questions about the relativities of local and national concerns, and it also raises questions about whether viewing these as binary oppositions is a useful way to measure impact.

Astronomy and place

In the search for suitable sites to host radio telescopes, the sparse habitation of the Karoo might easily give the appearance of 'emptiness' to outsiders. In fact, it is not uncommon for the sites of the SKA in Australia and South Africa to be referred to as 'remote' and 'desert', with the infrastructure that does exist in these places being viewed as a logistical advantage.[9] However, as this study has shown, the Karoo is not 'empty'; it has a deep history of human occupation adapted to its semi-arid ecology; and the community of Carnarvon and surrounding areas has a deeply contested past. Carnarvon, as a small but old-established town categorised as an agricultural service centre in development terms, is not in a strong position to negotiate against outside interests, a point well made by Atkinson (2019). Its local and provincial government structures are relatively new and weak, expertise is thin, the town itself is politically divided, and the Karoo is marginalised in South Africa's national spatial development policies. Carnarvon has high unemployment and existing resources are still concentrated in apartheid patterns; but it is also a stable place in terms of population, with little in- and out-migration. It is 'home' to most of its inhabitants, with the majority of its Coloured, Afrikaans-speaking occupants having been born in the area and having family links to the historical settlement of the place. To those living locally, the SKA's unfolding development could not have been anticipated, nor could it have been effectively resisted.

For the sake of perspective, it is worth comparing the SKA's South African site with that of its Australian counterpart, where contrasts are strong. The Australian SKA is located on a single very large cattle station (350,000 hectares) in Murchison, Western Australia, in an area with very few inhabitants. The Native Title of the traditional owners of the land, the Wadjarri Yamatji people, has been recognised through an Indigenous Land Use Agreement. Indigenous stories of the skies form part of the backdrop to radio astronomy's public face in Australia, and the telescopes have Wadjarri names.[10] In terms of the different interests of indigenous land use and astronomy more generally, it is also worth noting the long contest in Hawaii between first nations people and astronomers over the use of the sacred site of Mauna Kea (Hall, 2015) and, before that, contestations over sites in the USA.

In Carnarvon, the arrival of the SKA brought another wave of displacement to the historical patterns of land ownership. This time, I suggest, astronomy trumped the privileged position of white farmers in the historical patterns of possession and dispossession of the land. This huge shift was achieved incrementally and indirectly by policy moves after the political changes of 1994, rather than a single action of land redistribution. Two interrelated strands of policy enabled the changes to land ownership: policies in science development that elevated astronomy to a national priority with its interests capable of prevailing over others in designated places; and policies that elevated integrated environmental management above individual farming interests. In this set of moves, government regulations promoting the interests of science enabled the displacement not only of white farming interests, but also those of the communities whose interests had previously been displaced by white farmers and land enclosures – the hybrid community of Xhosa, Baster, and remnant San people who had gathered around the Rhenish mission at Schietfontein and whose descendants still live there. Astronomy, by means of government policies, provided the new 'colonising' logic of controlling the land and its use.

Astronomy policy shifts land ownership

The first step along this particular policy trajectory was the selection, in 2003, of the Karoo as South Africa's best site for large-scale optical and radio astronomy, considering its high atmospheric transparency, low levels of radio frequency interference, and low population density. The next step was the passage of the Astronomy Geographic Advantage Act (AGAA) of 2007, with the stated aims of preserving and protecting areas uniquely suited for optical and radio astronomy, providing for intergovernmental cooperation and public consultation on 'astronomy advantage areas', and setting out a framework to regulate future activities.

As a long-term instrument to protect astronomy, the AGAA made provision for extensive restrictions on land use in three types of astronomy advantage areas: core, central, and coordinated. And it signalled that all of the Northern Cape (except for its capital city district of Sol Plaatje) would be considered an astronomy advantage area. The Department of Science and Technology was recognised as the Astronomy Management Authority, and the Department entered a co-management agreement with the National Research Foundation (NRF).

In 2008, the NRF purchased two farms, Losberg and Meysdam, an area of approximately 13,500 hectares near the road between Carnarvon and Williston, to begin the construction of the KAT-7 as its proof-of-concept for the SKA-SA bid. From the perspective of astronomy interests, the fact that this was a long-established farming area was not an issue that warranted much consideration. To the outside eye, it was a sparsely populated semi-desert region of non-arable, low potential grazing land, whose contribution to the country's overall economy was relatively minor. In the words of the CSIR's impact study, 'The Karoo area consists of a low population density with isolated communities with fairly high levels of poverty and inequality which results in a vulnerable population' (2016, p. 18).

From the perspective of those living in and around Carnarvon, most with long-standing connections to the place as home, the initial purchase of the two farms was seemingly not of major concern.[11] Before long, however, rumours began to circulate that the SKA-SA was not being open with the local community about its future intentions and, more worryingly, that the organisation intended to purchase more land. Though SKA-SA representatives had held several meetings with locals and their municipalities in affected towns, community suspicions in Carnarvon grew. In a politically fragmented place with weak local and provincial government structures, the SKA-SA's consultative processes were beyond the cohesive capacity of the locals to engage with constructively.

In 2010, the MeerKAT site was formally declared as the Karoo Core Astronomy Advantage Area under the AGAA. This meant that the government would be able to acquire more land through purchase, and, if necessary, expropriation. It also meant the introduction of a number of restrictions, including restrictions on access to the core area, and on radio frequency use, light pollution, and aviation. In practical terms, it meant that farmers in the area within a radius of about 80 km from the SKA centre would be unable to use their cellular telephones – a move that generated bitterness in a community used to having its way. In the same year, part of the site was declared a National Key Point for high security protection.

Whereas in 2008 there had been some optimism that the radio astronomy venture might bring benefit to the town of Carnarvon and its surrounds – tourism, employment, infrastructure, and facilities – disillusionment set in, with the local community complaining that the SKA-SA's communications were less than frank, and that its activities would be disruptive to livelihoods in the area. Given that the SKA international bid process was still underway, the SKA-SA was no doubt operating with a measure of uncertainty about future plans. Nonetheless, it would also be fair to say that a good deal of secrecy surrounded its activities, and it is likely that engagements with local people would have generated greater expectations than the SKA-SA could meet (or had anticipated). Local people interviewed for this study voiced a number of complaints and uncertainties. They spoke of fears that Carnarvon would become a ghost town ('*spookdorp*'), with astronomy potentially draining future life from the existing community in its search for the distant origins of the universe. There were very few positive voices to be heard among those I spoke to. And indeed, resistance and disconnectedness continued in many of the local relationships with the SKA through the duration of my research.[12]

In 2014–2015, the government passed further regulations under the AGAA, setting out three zones of radio frequency spectrum use, this time for the Karoo Central Astronomy Advantage Area (CSIR, 2016, Chapter 1, p. 2). The SKA-SA began its land acquisition programme, and by the end of 2017, it had purchased 32 more farms around the MeerKAT core site, reaching negotiated prices with willing sellers and acquiring a further 107,000 hectares. Extending from the MeerKAT core, three spiral arms of mid-range instruments would be set up, stretching across 131 land parcels (an area of nearly 500,000 hectares) with servitude agreements (rather than purchases) to be arranged with land owners on matters such as access to roads, power lines, and land pockets, including restrictions on farming activities in corridors around spirals (Figures 9.1).

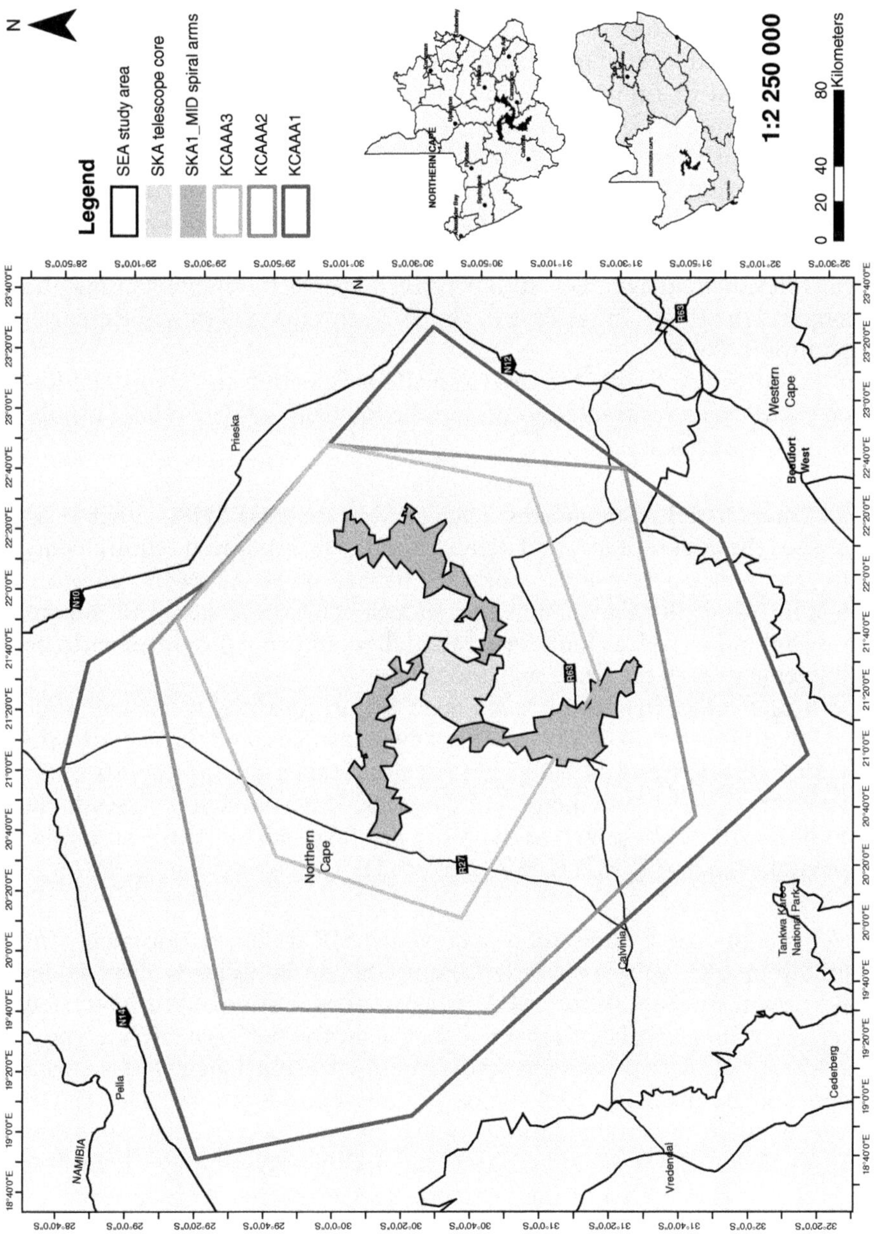

Figure 9.1 Northern Cape: SKA telescope array showing spiral arms and Karoo core astronomy advantage areas.

Environmental protection policy and land use

The MeerKAT and SKA-SA, as a large-scale venture of national and international significance, was identified as a Strategic Infrastructure Project (SIP) under the National Development Plan of 2012. SIPs are given special consideration with regard to planning, approvals, and implementation, and the Department of Environmental Affairs undertakes Strategic Environmental Assessments (SEAs) to address regulatory environmental requirements and ensure legislative compliance with the National Environmental Management Act. The SEA approach, intended to be a proactive process that brings a strategic emphasis to inform planning and continuous decision-making, produces an Integrated Environmental Management Plan (IEMP), with a relevant management authority responsible for implementing the outcomes of the Plan. (In the case of the SKA, this is the National Research Foundation/NRF.)

For the SKA-SA, South Africa's Council for Scientific and Industrial Research (CSIR) was contracted to undertake the SEA, and its report is published in two sections:

- an **Integrated Environmental Management Plan (IEMP)** which establishes the geographical and activity scope, the minimum requirements for the construction and operation phases of SKA1_MID, environmental principles, environmental management outcomes, and mitigation actions as well as long-term research monitoring programmes to be implemented on the SKA site; and
- a **Strategic Environmental Assessment Report** intended to provide supportive information on the process, key milestones, and the engagement with interested and affected parties undertaken during the Strategic Environmental Assessment. The Strategic Environmental Assessment Report will not be gazetted and does not form part of the legal implementation of the IEMP. (CSIR, 2016, p. 1)

The Astronomy Advantage Act as well as the SIP set out requirements for consultation, and the CSIR's reports, available on its website, provide extensive documentation on the SEA process, consultations with interested parties, as well as specialist reports for agriculture, heritage (archaeology, palaeontology, cultural heritage, and visual/landscape aspects), terrestrial ecology and biodiversity, and socioeconomic considerations. The IEMP sets out detailed recommendations on the construction of telescopes and measures taken to mitigate environmental and heritage impact. It recommends that the core area of the telescope be a Protected Area in terms of the National Environmental Management Act and anticipates that it will be declared a national park falling under the management of SANParks.

Looking at the details of the IEMP and its specialist reports in contrast to the issues raised by locals, what is striking is the different scales of their concern. In her research on Carnarvon, Butler (2018) evocatively identifies an embodied sense of place as part of the inner worlds of those born and raised there and suggests that 'the intimate scale of life in a small Karoo town is an important element in this sense of place' (p. 71). In the SEA documents, this intimate scale is evident in the concerns raised by locals, and contrasts powerfully with SKA-SA's more technical, abstract and large picture responses. Whereas the SKA-SA's focus in these documents is on environmental and heritage mitigation alongside the technicalities of telescope building, locals are more concerned that 'life here is ending as we know it'. A good number of the submissions showed that the nature of the SEA was not properly understood by locals, some of whom anticipated a more preliminary engagement.[13] A few examples will suffice to illustrate these points.

In relation to sheep farming, the SKA-SA points out that 'The Kareeberg accounts for 3.6% of sheep in the Northern Cape and 0.7% of the National sheep production', and therefore 'the impact in terms of sheep taken out of production for the area is 0.5% of the total sheep. This is less than 0.0192% of national sheep production'. For the farmers, these are *their* sheep, *their* local suppliers, and *their* abattoir that will be affected.[14]

Predator control is another issue that highlights different perspectives. For the SKA, this will be dealt with alongside other environmental measures; since there will be no sheep on the land, predator numbers will decline; and jackal-proof fencing will suffice around the site, rather than more costly game fences. For farmers, predator control is essential for stock protection, and predator hunting has become a ritual of its own in the area's communal life. Farmers' experience is that neighbours bear the repercussions if farms are left fallow, as illustrated also in the case of the town of Vanwyksvlei whose local economy nose-dived when farmland was lost to the Alkantpan missile testing range.[15]

From the SKA-SA's perspective, the detailed expert reports on agriculture, heritage, and environment illustrate technical capacity to understand and look after environment and heritage. In contrast, among farmers there is general disbelief that the SKA has the capability to look after this land or understand predator behaviour. What these examples illustrate is the incommensurability of the broad perspectives of large-scale infrastructure projects and the 'intimate scale' of daily life in a place. And, as Michael Gastrow's (2015) research shows, local voices are seldom given much space in media coverage about the SKA-SA's activities, which emphasises the politico-symbolic value of this prestige project for South Africa.[16]

Yet it would be mistaken to romanticise this local place or regard it as pristine and unsullied by change. Somewhat ironically, in SEA consultations farmers record their concerns for farmworkers and their families who would be displaced by the purchase of farms, most likely exaggerating their numbers. But many of the local farms had reduced their numbers of farmworkers in previous years, as mentioned in Chapter 6. The IEMP, stating its commitment to rehousing and retraining these workers, estimates the number of families affected at 24 – an average of less than one family per farm purchased.[17] Arguably, those directly affected by the SKA-SA's land purchases have received benefits in terms of payouts for farmers and retraining and rehousing for their workers. Whether or not benefits extend to the rest of the town is another matter. Certainly, this had been regarded as a marginal place before the SKA-SA moved in, with few prospects of investment under the National Spatial Development Framework and low potential for development.

The SKA-SA has reached memoranda of understanding with the farmers' organisation, AGRI-SA, and also with the San Council, a recently formed political group arguing for the recognition of San and their heritage. However, at the time of its launch, the SKA-SA had been unwilling to recognise or meet separately with the group of local people organised as the *SKA Belangegroepforum* (Concerned Group Forum), though its membership represents Coloured people who have deep historical roots to the land, anchored in the historical Rhenish Mission Church.[18] For the opening of the MeerKAT telescopes, a televised event was attended by dignitaries, and the SKA brought the San Council for a ceremonial blessing of the land – San people living some distance away and in other parts of the country flown in for the event. But there was no equivalent recognition of the local community who were displaced by white farmers in the 1800s. That this community would most certainly have San descendants quite apart from any San Council is argued by Parkington, Morris and de Prada-Samper (2019). These authors claim from archaeological evidence that 'By the mid 19th century, small towns across the Karoo completed the dispossession and transformed residual hunters and gatherers into farm labourers, domestic servants and marginal town-dwellers with little or no political or social coherence' (pp. 1–2). They continue as follows:

> |Xam culture persists, albeit under altered, arguably reduced, circumstances, articulated through the Afrikaans not the |Xam language, maintained by scattered, residual communities of largely Christian faith and deprived of its hunting and gathering logic. Nevertheless, |Xam descendance is a detectable and increasingly claimed inheritance, one that has survived nearly 300 years of colonialism, empire, apartheid and post-apartheid discrimination. How, we ask, is social justice best delivered to such people and communities?
>
> (2019, p. 2)

While the San Council is a body of some significance in terms of political recognition, the question remains as to whether or not it has a local presence relevant to the community in Carnarvon. Given this question, the SKA-SA's gestures to the San Council at its opening came across as empty symbols to this community. In my own view, it would be a mistake to regard the SKA-SA's gestures of 'indigenous recognition' as being parallel in any substantive way to the recognition given to Indigenous land rights on the Australian SKA site. To see them as similar would be risible.

In terms of local interests in relation to those of big science and the SKA, the following comment recorded in the CSIR's response trail presents the tension clearly:

> It is quite possible that at the end of the study the findings of the socioeconomic [survey] may be that there few benefits for the local community but we all need to look at the bigger picture. It is a big project. It has advantages for South Africa as a country. It makes South Africa one of the best scientific destinations in the world. We do a lot of work with Home Affairs and some of the best radio telescope scientists come to South Africa and establish themselves here. That means that we start to build up a knowledge economy. Coal for example will eventually run out, and the economy has to move away from a resource rich economy to a knowledge economy and the SKA is able to influence this.
>
> (n.d., p. 48)

What will remain of the town of Carnarvon after the loss of productive farms and the flow-on consequences for the local economy is an open question. The relationship between local and national interests is a complex and continuing tension.

Development questions: 'big science' and basic provision

Both in its reports to parliament and in engagement with the public, the SKA-SA (now part of the South African Radio Astronomy Organisation (SARAO) under the NRF) takes the position is that it is merely a 'project', not the government, and cannot be expected to deliver social services that government departments are responsible for. To do so would be to take focus and funding away from the major science project of astronomy, which is massive as it is, and is delivering to target.

While there is a logic to this argument on the scale of the project itself, taking this stand is perhaps a little disingenuous in that distancing the SKA-SA from the government also deflects responsibility for the government's decisions to invest in 'big science' when faced with other massive development needs in a fundamentally unequal society. It could be argued that if interest groups could work with the government to secure inter-departmental cooperation to fund the SKA, there is no reason why this could not be secured to take care of the local effects of the project – if this were regarded as worthwhile or if big science were seriously linked to local development in government priorities. As Atkinson (2019) points out, the SKA processes have not involved district or local government bodies in any substantive way, although national policies on spatial development planning state the government's responsibility for capacity-building. In her view, 'The establishment of the SKA within the Science and Technology sector provides a classic example of a centralist government department negotiating its way, on its own terms, through a multi-polar institutional maze' (2019, p. 690). In the face of this, regional and local institutions have been left 'bemused' and 'passive'. In effect, what this shows is that where local communities have weak structures and are politically divided, it is not possible for them to engage with a national project of this scale in anything but a nominal way.

President Ramaphosa's State of the Nation Address of 2019 continues the rhetorical link between big science and service provision:

> As a young nation, only 25 years into our democracy, we are faced with a stark choice. It is a choice between being overtaken by technological change or harnessing it to serve our developmental aspirations. It is a choice between entrenching inequality or creating shared prosperity through innovation. Unless we adapt, unless we understand the nature of the profound change that is reshaping our world, and unless we readily embrace the opportunities it presents, the promise of our nation's birth will forever remain unfulfilled.
>
> Today, we choose to be a nation that is reaching into the future. In doing so, we are building on a platform of extraordinary scientific achievement. The successful construction in the Northern Cape of the MeerKAT telescope, the world's largest and most sensitive radio telescope, and the development of the Square Kilometre Array has enabled South Africa to develop capabilities in areas such as space observation, advanced engineering and supercomputing. These skills and capabilities are being used to build HERA, a radio telescope designed to detect, for the first time, the distinctive radio signal from the very first stars and galaxies that formed early in the life of the universe.
>
> This is not merely about advancing human understanding of the origins of the universe – it is about responding to the challenges that face South Africans now and into the future. It is about developing the technology and the capabilities that will build a dynamic and competitive economy that creates decent, sustainable jobs. It is about enhanced food security, better disease management, and cheaper, cleaner and more efficient energy. It is about smart human settlements and social development solutions built around people's needs and preferences.
>
> It is about smarter, more responsive, more effective governance.[19]

However, it remains the case that South Africa is profoundly unequal, service provision is a source of continuing dissatisfaction at the local government level, and the country's basic education system fails the majority of young people. Beyond visionary statements, the huge investment of government resources in science has uncertain links to meeting social and development needs at this point, and this debate should not be glossed over or sidestepped. Nor should it be set out in binary terms as a stark choice, rather than engaged with in constructive ways. Given the enormous resources involved, it is surely not beyond the remit of the government to ensure greater inter-departmental oversight of practical steps to address developmental initiatives to link SKA-SA expenditure to tangible outcomes – or to end the rhetoric.

The SKA and schools in Carnarvon

The umbrella body for the SKA-SA, SARAO, includes among its activities a separate project on Human Capital Development (HCD), which was started in 2005. SARAO reports regularly to parliament on its achievements in building capacity in science and technology through bursaries for undergraduate and postgraduate students, establishment of funded research chairs, research grants of various sorts, and technician training (particularly to meet its own maintenance and development needs). It also publicises its human capital and development initiatives in the Northern Cape in brochures on its website.

Reporting to parliament in September 2019, the HCD achievements in building academic and technical capacity were summed up as follows:

- Total of 1166 grants and scholarships awarded, of which 607 have been awarded to black South Africans, and 275 to South African women.
- Higher than average graduation rates: 76% for undergraduate, 97% for honours, 94% for masters, and 92% for doctoral levels.
- Employment of HCD postgraduates: 24% by SARAO, 37% by RSA universities and national facilities, and 19% into RSA high-tech industry.
- To date 84 students have been trained as electricians, fitters, and turners; in instrumentation, diesel mechanics; in IT and boiler making; as well as in carpentry, plumbing, bricklaying, and welding in a technical training centre in Carnarvon.
- Of 21 graduates who completed an extended (transformation) 3-year internship programme for science and engineering graduates, 16 have been employed in permanent positions in SARAO, and 5 have taken up positions in industry.
- At the school level, 181 bursaries have been awarded for students to attend *Hoërskool* Carnarvon, and 15 matriculants from the school have been awarded undergraduate bursaries (though no information is given about the success rates of these school-level programmes).[20]

This is an impressive record of building capacity in science and engineering at a national level, and the HCD programme is path-breaking in South Africa in this regard – and most likely a record for similar countries. From the perspective of this study, it illustrates is that it is possible to make considerable and continuing progress on a national level, particularly in post-school programmes where matriculants with good results are selected to study science and engineering or develop technical skills. However, interventions to improve opportunities at the school level are more complex, particularly in contexts where schools are in conditions of poverty and structural disadvantage, and education departments are weak.

Indeed, it could be argued that the SKA's engagement with schooling in the Karoo towns, including Carnarvon, illustrates some of the difficulties of the schooling system as a whole. Much as the SKA's interventions are intended to bring benefit to local schools, they run up against, and expose, fundamental weaknesses in the schooling system itself. To illustrate this point, the SKA-SA publicity on its websites offers a place to start, relating to the supply of equipment, additional staff, and bursaries. The website publicity includes the following statements about its engagements with schools:

> SKA SA has used its Schools Programme to help to recruit qualified maths and science teachers, construct and equip a Cyberlab and science laboratories, provide role modelling and career guidance, and fund field trips and exchange programmes.[21]

> The South Africa project also supports the schools in the Karoo area, where the SKA will be sited, and has given bursaries for students from surrounding towns to study mathematics and science in the town of Carnarvon. A Community Knowledge Centre and other projects have been built to uplift the community.[22]

With regard to equipment, cyberlabs and science laboratories are important for all schools and there is no doubt that making these available to Karoo schools is beneficial. This is particularly the case in *Hoërskool* Carnarvon, which as a secondary school enrols students from the primary schools in surrounding towns. Students have been introduced to robotics, and this provides possibilities for competitions and excursions out of town. In terms of the curriculum, functioning laboratories enable greater engagement with science as a subject. That said, there is always a frustrating gap between the provision of equipment and how it is used, the former being much more amenable to intervention than the latter. Crucially, teachers need to be able and willing to use technologies. Computers easily lie unused or underused and often cannot be repaired if they malfunction. Continuing support is always needed beyond installation and remote places are harder to monitor and assist. My observations in Carnarvon's schools suggest that they are not exceptions with regard to technology implementation.[23] Research by van der Hoef (2017) on the actual use of the Community Computer Centre in its early stages shows that facilities are always one step away from dysfunction in communities that are remote and under-served. While it is indeed better that the facilities are provided rather than not, the SKA's publicity documents seem over-ambitious in tone.

With regard to the recruitment of qualified teachers, the principals of both schools in Carnarvon view this as one of their major challenges, given the poverty and remoteness of the Karoo region as well as the Afrikaans medium of the Carnarvon schools. It is clear from my research in Carnarvon and more widely that the benefits of adding an extra staff member (as in the government's Dinaledi project) depend almost entirely on the quality of the staff member added.[24] For the duration of the present study, the SKA-SA used *Teach South Africa* to supplement the staff of Carnarvon schools – young graduates without professional teaching qualifications, who were visited by an expert educator for support. Anecdotal observations from my research study suggest some of the difficulties of this approach to teacher recruitment. Again, much depends on the suitability of the appointee for the position. The physics graduate appointed to *Hoërskool* Carnarvon high school (who is mentioned by name on the SKA-SA website) was a successful teacher who could speak Afrikaans and the results he achieved were good. (He did, however, leave the school to pursue other interests, though he later returned.) By contrast, the mathematics graduate recruited alongside him could not speak Afrikaans, and consequently found classroom practice challenging in her first year at the school even though her content knowledge was good. In the second year of her appointment, she was tasked with teaching English (not her expertise, or even her first language). The practice of teachers being assigned to classes outside their expertise is a documented problem of teacher deployment in South African schools. In the primary school, experienced teachers raised concerns in interviews about the professionalism of appointees who were unwilling to accept guidance from them. On one of my visits, accompanying the principal to a classroom where a young appointee wearing a baseball cap backwards and chewing gum arrived late for class, I was told 'At least he's local and we know him'. Both school principals spoke of the difficulties of attracting and retaining good young teachers to schools in remote locations such as this.

The SKA-SA's bursary programme offers opportunities for primary school children in surrounding towns to attend secondary school in Carnarvon, and for secondary students to attend universities. As an individual-level intervention, bursaries provide valuable opportunities for those who have aptitude for mathematics and science, and in the case of Carnarvon, they have opened pathways for individuals that were not there before. However, they cannot address the structural disadvantage that prevails in schools such as these, which have large classes that include many repeating students who were 'progressed' without meeting the curriculum requirements for the grade (as discussed in Chapter 8). A number of interviewees thought the emphasis on mathematics and science in these bursaries was too narrow for the schools' needs, and possibly distorting of the broader curriculum. Concerns were raised by a number of interviewees (including by a district official) about how to provide support for children who had been awarded bursaries but failed and been sent away.

Arguably, these points of criticism reflect the perspectives of people who are not able to benefit from bursaries.

My intention in making these points is not to disparage the SKA-SA's efforts in local schools, so much as to caution against overblown media publicity about social upliftment. I suggest that more careful arguments and interventions are necessary if the benefits of big science are to connect with actual practices in schooling to make a significant difference to more than a few individuals.[25]

Conclusion

The intention of this chapter has been to illustrate the contrasting scales between the SKA as a global and national project, and a marginal local place in order to show the multiple scales of activity and their different logics in a single 'fragment of space'. The snapshot provided here freezes a moment in time in a changing picture, but it nonetheless enables different perspectives to be viewed. It also enables a different variant of the historical patterns of possession and dispossession to be traced, in this case, from white farmers to the SKA-SA and its telescopes. From the perspective of the SKA-SA, the project of building the world's largest radio telescope in this place is a massive set of activities with global reach that extends decades into the future, and it has achieved remarkable success so far in meeting its goals. A contrasting local perspective is provided by the Coloured community gathered around the original Rhenish Church, later the NGK *Sendingkerk* and currently the Uniting Reform Church in Carnarvon. Perhaps this historical community was simply too hopeful, and its expectations of the SKA too ambitious. Nonetheless, the impact of the SKA on the future is likely to be very different for them, as illustrated by the following address by Dr Izak (Sakkie) Potgieter to his congregation:

> The elephant that approached on the horizon was so fascinatingly beautiful. Progress was at last approaching in the name of science. The square kilometre array (SKA) would take us, this insignificant oppressed and downtrodden people, onto the next level of existence. Justice will at last prevail. The oppressors will be trampled upon the way the escaped convict from the Cape Colony, Karel, was trampled upon towards the end of the nineteenth century, ensuring his eternal fame with the naming of the farm, Karelsgraf.
>
> The farms that were stolen from their rightful owners would at last be returned – Minaskolk, Bruinheuwel, Rooileegte, Skurwekloof, Konka, Kokerboom, Swartbaadjie, Boezak, Waaipunt, Bloubos, Rooi Uitspanning, Didas, Xqinika, Koerieskloof, Berge (a collection of five farms). In this way the memories of the proud and honest owners, Paulus Vass, Jan Alfred Janes, Gert Petoors, 'Peacock' de Bruin, Jan Haas, Esau Hoorn, the brothers Jan and Hans van Wyk, Flip Links,

Jongie Lamoela, Dawid Julius, Dawid Jann, Abraham Andreas Jona Malgas, and Sikumbini would be honoured.

The old generation who followed Ouma Katriena Haas in 1997 in singing the sad Dutch song, would once again sing this song in triumph:

> Ons, Bruinmense
> die seuns van die slawe vra onse eie land terug
> wat gesteel was van onse vaders
> toe hulle in die donker sijn.
> Gee dit terug nou, gee dit terug nou
> vir Suid-Afrika.
>
> *[Translated freely:*
> *We, Coloured people,*
> *the sons of slaves ask our land to be returned,*
> *which was stolen from our fathers*
> *while they were in the dark.*
> *Give it back now, give it back*
> *to South Africa.]*

The elephant arrived and started trampling on them, the oppressors. But of course an elephant could not distinguish between the oppressor and the wretched of the earth. It is merely an innocent animal, for God's sake. It also started trampling on us, the people who chose to forgive, to choose a life of peace above one of hatred and revenge.

What will I write up in this, my final eulogy to the true heroes of Carnarvon?

Let us be brave, brothers and sisters, and bear the consequences of 'progress'. We have no choice.

(Potgieter, personal email communication, 2017)

Notes

1 I have made no attempt to provide new insights through primary research since the complexity of the SKA could overwhelm the focus of this study on the town and its schools. Instead I have relied on secondary sources, material available on the SKA's website, and its reports to parliament.

2 From interviews and census data, I gather that most of the people who live in Carnarvon were born there or in the northern Cape, and many people 'orbit' between a set of towns including Upington, Kimberley, de Aar, Oudtshoorn and Calvina, with Victoria West and Vanwyksvlei as near neighbours.

3 For a discussion on 'big science' and the SKA, see Gastrow and Oppelt (2018). They note

> The notion of "big science" has emerged as shorthand for the increasingly large science projects that first proliferated during the Cold War. The term was coined in 1961 to refer to a post-war political economy in which scientific research was a national security priority requiring state intervention and resources....
>
> (p. 1)

See also Gastrow, Kruss, and Petersen (2016, pp. 361–375) and the special issue of *Journal of Southern African Studies* (2019).

4 Wild notes that the SKA is comparable in scope to the European Organisation for Nuclear Research (CERN) and the Large Hadron Collider.

5 MeerKat project cost in total about R2.2 billion https://pmg.org.za/committee-meeting/24534/. Wild (2012, p. 89) quoting Fanaroff: 'Megaprojects have a way of obtaining funding from different buckets'.

6 These countries were Botswana, Ghana, Kenya, Madagascar, Mauritius, Mozambique, Namibia, and Zambia.

7 Notable among SKA supporters were Rob Adam and Bernie Fanaroff, supported by strong ministers such as Naledi Pandor (Science and Technology), Trevor Manuel and Pravin Gordan (Finance), and Alec Erwin (Public Enterprises). See Dubow (2018, p. 23); Wild (2012, p. 90).

8 Walker and Chinigò (2018, p. 12). See also Kahn (2008):

> Joining the Big Science bandwagon brings the benefit of "feel good" through the possibility of working with the community active in frontier science. How such activity would address the needs of emergent industry or social disparities is yet to be explained. One fears that a trickle-down mechanism is implicitly at play. (p. 10)

There are a number of relevant articles in the Karoo Special Issue of the *African Journal of Range and Forage Science* themed Trajectories of Change in the Anthropocene.

9 See Chrysostomou, Bolton and Davis (2018),

> The Murchison Radio Observatory in Western Australia and the Karoo Radio-Astronomy Reserve are two of the most remote locations on the Earth, ensuring minimal impact from RFI. A consequence of this remoteness is that providing power to these locations is a big logistical challenge, and as such any infrastructure that is present or requires installation, is purely for the operation of the telescopes themselves.

10 Sim (2014):

> In 2009 the Yamatji Marlpa Aboriginal Corporation, which is the legal representative group for the Wajarri Yamatji claimants, along with CSIRO and the Australian and Western Australian governments finalised an Indigenous Land Use Agreement (ILUA) that gave consent for the early radio astronomy. The ILUA is providing employment, educational and mentoring opportunities for the Wajarri Yamatji people, as well as financial support for the community.

Among the first names are Bundarra (stars); Wilara (the Moon); Jirdilungu (the Milky Way).

11 The points made about local community responses are drawn from observations and interviews in my research in Carnarvon, together with the CSIR (2016) reports which include response chains. I have taken care to substantiate the evidence for the points I make here, given their sensitive nature. See also Potgieter (forthcoming).

12 That said, my study provides only one snapshot of this time/space and no doubt the dynamic of encounter would have shifted on since then in ways not necessarily possible to predict.

13 Extensive information about the CSIR's reports on the SKA, including briefing documents and comments and response trails are available on their website http://www.skaphase1.csir.co.za/

14 CSIR (2016). See also Kirsten (2016).

15 Kirsten (2016):

> Alkantpan is an area equivalent to 85 000 hectares acquired in 1987 by the South African Defence Force for a missile and ammunition testing terrain. After a few years farmers on the neighbouring farms found it very hard to farm profitably with small stock due to high losses of livestock caused by the rapid increase in the predator population on Alkantpan... Eventually many farmers left their farms because it has become impossible to farm. No management plan and no grazing plan have proven to be successful to prevent these huge losses.
>
> (p. 15)

16 Gastrow (2015, pp. 703–722).
17 http://www.skaphase1.csir.co.za/wp-content/uploads/2017/07/Comments-and-Responses-Trail_IEMP.pdf.
18 The group has a Facebook page, https://www.facebook.com/kreebergfees. There is a WhatsApp group, and the local forum has written records which I have seen.
19 https://www.gov.za/speeches/president-cyril-ramaphosa-2019-state-nation-address-7-feb-2019-0000.
20 SKA briefing to Parliamentary Portfolio Committee, 11 September 2019.
21 http://www.ska.ac.za/ska-activities-in-the-northern-cape/nurturing-learners-talent/Accessed 6 August 2019.
22 https://southafrica.skatelescope.org/about/ Accessed 6 August 2019.
23 These points accord with my experience of schools and technology elsewhere, and reflect my interviews, informal observations, and discussions in both the primary and high schools in Carnarvon. And indeed, in my view it would be surprising if the SKA's HCD programme did not reflect limitations such as these.
24 Staffing of schools in non-metropolitan places produces a set of problems of their own. People with formal qualifications may lack classroom competencies and indeed may be 'parked off' in schools that struggle to attract good staff for reasons of location. Suitably qualified teachers and deputy principals may take up positions they are qualified for but not really interested in for personal reasons (such as moving with a partner) – and in the process may block others who are genuinely interested.
25 'Distortion' includes the irony of a girl living in a shack winning an international prize for an assignment on Carnarvon and tourism – when in fact the SKA does not intend to generate tourism at all.

References

Atkinson, D. (2019). When stars collide: Competing development paradigms in the central Karoo. *Journal of Southern African Studies*, 45(4), 689–709.

Butler, S. (2018). Knowledge relativity: Carnarvon residents' and SKA personnel's conceptions of the SKA's scientific and development endeavours. MA thesis, University of Stellenbosch.

Chrysostomou, A., Bolton, R., & Davis, G. R. (2018). The Square Kilometre Array: Challenges of distributed operations and big data rates. *Proc. SPIE 10704, Observatory Operations: Strategies, Processes, and Systems VII*, 1070419 (10 July 2018). doi: 10.1117/12.2309554.

Council for Scientific and Industrial Research (CSIR). (2016) Integrated environmental management plan for the South African mid-frequency array of SKA Phase 1. CSIR Report Number: CSIR/02100/EMS/ER/2016/15241/B. (Second version 2018).

Council for Scientific and Industrial Research (CSIR). (2017) Second comments and responses trail for the Strategic environmental assessment of the SKA mid-frequency array of SKA Phase 1, Report Number: CSIR/IU/021MH/ER/2017/0008/B.

Council for Scientific and Industrial Research (CSIR) (n.d.). Comments and responses trail for the strategic environmental assessment of the South African mid-frequency array of SKA Phase 1 (SKA1_MID). http://www.skaphase1.csir.co.za/wp-content/uploads/2017/01/Comments-and-Responses-Trail.pdf Accessed 26 November 2019.

Dubow, S. (2018). 200 years of astronomy in South Africa: From the royal observatory to the 'Big Bang' of the Square Kilometre Array. *Journal of Southern African Studies*, 24–25. doi: 10.1080/03057070.2018.1496700.

Gastrow, M. (2015). Science and the social media in an African context: The case of the Square Kilometre Array telescope. *Science Communication*, 37(6), 703–722.

Gastrow, M., Kruss, G., & Petersen, I. (2016). Connecting capabilities in highly unequal developing countries: The case of the Square Kilometre Array telescope in South Africa. *Development Southern Africa*, *33*(3), 361–375.

Gastrow, M., & Oppelt, T. (2018). Big science and human development – What is the connection? *South African Journal of Science*, *114*(11/12), Art. #5182.

Hall, S. (2015) Hawaii's telescope controversy is the latest in a long history of land-ownership battles. *Scientific American*, 11 December. https://www.scientificamerican.com/article/hawaii-s-telescope-controversy-is-the-latest-in-a-long-history-of-land-ownership-battles/ Accessed 26 November 2019.

Kahn, M. (2008). The contract between science and society: A South African case study. *Science and Public Policy*, *46*(1), 1–10.

Kirsten, J. (2016). An estimation of the agricultural economic and local economic impact of Phase 1 of the SKA. Background paper prepared for the CSIR.

Parkington, J., Morris, D., & de Prada-Samper, J. M. (2019). Elusive identities: Karoo |Xam descendants and the Square Kilometre Array. *Journal of Southern African Studies*. doi: 10.1080/03057070.2019.1647655.

Potgieter, I. C. (forthcoming). The SKA in Carnarvon. Decoloniality and modernity in a Calvinist context.

Republic of South Africa. (1998). National Environmental Management Act. Act No. 107 of 1998. Pretoria: Government Printers.

Republic of South Africa. (1996). *White Paper on Science and Technology: Preparing for the 21st Century*. Pretoria: Government Printers.

Republic of South Africa. (2007). Astronomy Advantage Act. Act no. 21 of 2007. Pretoria: Government Printers.

Sim, H. (2014). The kingdom of quietness. *Issues*. http://www.issuesmagazine.com.au/article/issue-march-2014/kingdom-quietness.html.

Van der Hoef, M. (2017). The local, the global and the self: An ethnographic account of a community computer centre in Carnarvon, Northern Cape, and its significance for its users' sense of self and their place in the world. MA thesis, University of Stellenbosch.

Walker, C., & Chinigò, D. (2018). Disassembling the Square Kilometre Array: Astronomy and development in South Africa. *Third World Quarterly*. doi: 10.1080/01436597.2018.1447374.

Whitelock, P. A. (2004). Optical astronomy in post-apartheid South Africa: 1994 to 2004. In A. Heck (Ed.), *Organisations and Strategies in Astronomy*, *5*, 39–60. Kluwer Academic Publishers.

Wild, S. (2012). *Searching African Skies: The Square Kilometre Array and South Africa's quest to hear the songs of the stars*. Johannesburg: Jacana.

10 Towards decolonising schooling

Realising the impossible dream?

The decolonial moment

The dismantling of apartheid, which began in 1990, signalled the end of an era of colonialism, which had significance well beyond South Africa.

Thirty years before this, in February 1960, British Prime Minister Harold McMillan had signalled the end of empire in a speech delivered to the white parliament of South Africa, which was then a British dominion. 'The wind of change is blowing through this continent', he solemnly observed, 'And whether we like it or not this growth of national consciousness is a political fact'. McMillan also voiced his concern that the 'achievements of western civilisation' – the 'freedom, order and justice' of its way of life – would be tossed aside by nationalist movements seeking to rupture ties with their colonisers. It is hardly surprising, then, that McMillan, as a guest of the Verwoerd apartheid government, did not accept the ANC's request for a meeting during his visit.

At the same time as McMillan's valedictory to empire, Franz Fanon (1961/1990) published his excoriating critique of colonialism in *The Wretched of the Earth*, writing from Algeria, at the other end of the African continent. Fanon declared in no uncertain terms that colonised peoples experienced 'western civilisation' in quite different terms to the 'freedom, order and justice' set out by McMillan. In the compartmentalised world of the colony, colonised people's experience was of violence and forced labour; plunder and pillage; dispossession, bodily degradation, and denial of their humanity – an experience later named 'the dark side of modernity' by theorists of decoloniality. Fanon was contemptuously clear that colonised people should not be expected to embrace the 'way of life' so positively projected by colonisers:

> The violence with which the supremacy of white values is affirmed and the aggressiveness which has permeated the victory of these values over the ways of life and of thought of the native mean that, in revenge, the native laughs in mockery when Western values are mentioned in front of him ... In the period of decolonisation, the colonised masses mock at these very values, insult them and vomit them up.
>
> (1961/1990, pp. 33–34)

Fanon also warned against inherent weaknesses in colonial national consciousness – pitfalls by which it could become 'an empty shell, a crude and fragile tragedy of what it might have been'. The weakness, he argued,

> is not solely the result of the mutilation of the colonised people by the colonial regime. It is also the result of the intellectual laziness of the national middle class, of its spiritual penury, and of the profoundly cosmopolitan mould that its mind is set in.
>
> (1961/1990, p. 119)

As it happened, the South African government shrugged aside McMillan's observations and battled against the 'wind of change' for a further three decades. It chose to leave the British Commonwealth as a republic, to tighten apartheid's grip and uphold white supremacy for as long as it could, and to violently suppress national consciousness other than its own. Over the decades, the project of decolonisation became increasingly more complex even as it became increasingly more likely.

Part of the complexity faced by South Africa's new democratic government was its timing on the global stage: the Cold War that McMillan feared was at its end by 1990, the Berlin Wall had been torn down, and neoliberal capitalism was globally ascendant. With the collapse of a socialist utopian vision that had inspired the liberation movement, the challenges of entering the global marketplace on neoliberal terms, and the compromises of a negotiated settlement, the new government operated on unsteady and unfamiliar terrain. To this complexity on a global scale would soon be added the movements of migrants and the crisis of climate change, an awareness that the earth needed to be considered in its own terms beyond human exploitation, and the health pandemics of HIV/Aids and later COVID-19. In short, the timing of the victory of South Africa's national liberation struggle meant that the legacy of colonialism needed to be addressed in complex and rapidly changing global circumstances, different from the 1950s context in which the Freedom Charter had been written. Establishing a modernist constitutional democracy and equal citizenship in these conditions has produced wicked socio-economic and political problems for South Africa, as this book has shown, and these conditions pose challenges on the terrain of ethics and ontology as well.

As earlier chapters have illustrated, the dismantling of apartheid was uneasily achieved on the basis of tough negotiations, with much future work to be done. The intense powerplays, vested interests, and competing ambitions of the negotiated settlement produced uneven outcomes. Unresolved issues – particularly relating to land and to traditional authorities – remained to be settled under the 1996 Constitution. The contradiction between political freedoms and limited economic and social change is evident in the continuing poverty, inequality, and unemployment inflected by race and gender, and in continuing social restlessness.

Fanon's work has been one of the reference points for theorists of de/coloniality, concerned to understand the multiple and intertwined power relations that sustain colonial inequalities beyond the moment of governmental change. Although present-day South Africa cannot be read through the lens of Algeria in the 1960s, Fanon's insights on the ravages of colonialism and the mammoth task of forming new societies in its wake still have relevance. Of particular salience are his cautionary observations about the weaknesses of national middle classes who take over from colonial regimes. Without having wealth of their own, the national bourgeoisie could become the intermediaries to big capital; taking over the positions of former colonisers, they could be unable to steer change or set the national economy on a new footing.[1] That said, Fanon's message was one for humanity also: he argued that national consciousness would need to give way to political and social consciousness – in his view, a different form of humanism – for liberation to bring betterment to the lives of the majority. National governments would need to throw all their efforts into building an inclusive economic programme in order to fight poverty, ignorance, and injustice, and to give dignity to all citizens. This is a point I return to later.

This book has told the story of Carnarvon and its schools in two time periods to show the establishment of colonial relationships in the 1800s and the process of their dismantling after 1990. Drawing on Lefebvre's notion that the analysis of a 'fragment of space' may potentially disclose 'not just one social relationship but a host of them' (1991, p. 88), I sought to trace rhythms of practice in this place on different scales and at different times. With reference to broader analyses of policy and history of schooling, the study has magnified one place in two time periods to provide a more granular description of relationships, activities, and meanings than is possible in more general accounts. In studying the larger picture from the vantage point of a particular place, it has traced a complex range of issues: how power relations and racial hierarchies are established and maintained; how interest groups gain and maintain their privileged positions – while others oppose them; how human agency is able to shift government intentions in local contexts; how social imaginaries inform actions in ways that link the present to the past and future; the uneasy relationship between 'big science' and social development; and how schooling folds into broader intersectional power relations, social interests, and imaginaries. There is no single story of schooling, colonialism, and decolonisation, and the story of Carnarvon is illustrative without necessarily being typical. Nonetheless, when viewed alongside broader accounts of schooling and social change, this study of a specific place provides insights into the multiple ways in which inequalities in schooling take shape, endure, and shift, and what might be required to work towards justice in current times of global complexity.

In this chapter, I reflect further on questions of decolonising schooling. In particular, I return to questions raised in Chapter 1: how to understand the intersectional and entangled inequalities running from colonial to current times, how to work epistemologically within border contexts rather than asserting simple universalisms, and how to shift the form of current conversations towards a different ethics of engagement on how to live together with all others in the world we share – an all-world ethics.

Entrenched inequalities of coloniality and the illusion of sameness

It is not hard to describe the inequalities embedded in South African education from colonial and apartheid days, through post-apartheid policy shifts and into the present – though explaining these inequalities is the subject of continuing debate. As set out in Chapter 2, there is general agreement that the education system as a whole performs dismally, with South Africa ranked among the worst performers on all international comparative scales.[2] There is also general agreement that performance patterns in the education system are bimodal: there are distinctively different results for students attending different schools, and these results differ according to the poverty quintiles and former apartheid departments of schools. In the bimodal results, nearly 80% of students attend the poorly functioning part of the system, with a small minority (8%) attending the fee-paying schools (mostly desegregated) that achieve good results.[3] Almost all of the poorly performing schools are black schools in rural areas and townships. These patterns of performance have persisted in spite of post-apartheid policies which aim to achieve greater equity and quality in education.

In the face of evidence of poor performance, there have been numerous interventions and proposals for change, targeting different aspects of the schooling system seen to be problematic. Interventions target the 'usual suspects' for school reform: teachers (accused of being not sufficiently knowledgeable or not covering the curriculum); school leadership and management (accused of not running the school effectively as an organisation, or not adequately monitoring curriculum planning and resource use); parents (accused of not reading enough to their children or not being sufficiently supportive of schools); students (accused of not attending regularly or not working hard enough); and so on. Seldom are the policies themselves or the design of the system included in exercises of problem identification, save for criticisms of some of the provincial departments that are earmarked for inefficiencies and corruption.

Exploring the failure of most interventions to change what happens in schools, Larry Cuban (1988) points out that the basic organisational structures of western schooling have hardly changed over more than a hundred years. Students are sorted by age and grade (which he calls 'chunks' of curriculum); teachers work alone in separate classrooms; and each school has its own principal. This basic structure has endured in the face of numerous attempts at school improvement and school change, and this structure has been the template for colonial schooling as well. Explaining the endurance of schools in the face of interventions to change them, Cuban applies the distinction between first-order and second-order changes. First-order changes are changes of quality control, aiming to improve what exists while leaving the basic organisational form intact. They may address any part of the schooling system, including the range of issues mentioned above, and may be valuable in ameliorating problems. By contrast, second-order changes are changes in system design, targeting the basic organisational form of schooling. Second-order changes, Cuban argues, are much less likely to be successful than first-order changes, not least because their success depends on social and political changes occurring outside of schools.

Using Cuban's heuristic, it could be said that the major social and political changes of the 1990s provided enabling conditions for second-order changes to be made to South African schooling, and indeed, changes were made to its basic organisational design. Through the National Education Policy Act (NEPA) and South African Schools Act (SASA), racially based departments were collapsed and replaced with new national and provincial departments; a strongly decentralised form of school-based management was introduced; and a different funding model was put in place, introducing fees to supplement state allocations. In addition, provision was made for a relatively small portion of the budget to be allocated on equity principles. Curriculum 2005 may also be viewed as a second-order redesign to replace the apartheid curriculum with outcomes-based education (a redesign, followed by a number of quality control revisions before the design was abandoned).

Questions may be asked, then, about the adequacy of the system redesign in relation to its stated goals of 'redressing past injustices in educational provision' and 'providing an education progressively high quality for all' (as set out in the Preamble to the South African Schools Act). Is the system as redesigned capable of redressing past injustices? Will all public schools be able to progress towards the same quality of education provided in the well-performing parts of the system – or will 'high quality' mean different things for different parts of the public system?

I suggest that there are limits to what the present system can achieve, and that its basic design has limited capacity to redress apartheid injustices, overcome the deep inequalities between schools, or provide equal quality for all in terms of experiences and outcomes.

The major political changes of 1994 offered a moment for fundamental change – a decolonising moment. However, the education policy settlement of the 1990s was the product of broader negotiations, and in these negotiations the apartheid National Party – determined to secure the future of Afrikaans-medium schools – succeeded in ring-fencing former white schools and achieving a privileged status for them all.[4] The newly formed national education department, weak from the start in relation to schooling, was increasingly disconnected from the actually existing conditions in schools as it drew up its policies in ways that conformed to modernist parliamentary governmentality – white papers, national commissions, portfolio committees, and so forth. Policy formulation was ongoingly top-down in mandating change.[5] The division of powers between national and provincial departments hampered efforts of change in cumbersome bureaucracy, and possibilities for applying equity measures were further impeded by the allocation of extensive powers to the level of school governing bodies – schools that reflect, albeit with some changes, the profiles of their apartheid histories. Changes in curriculum did not take into account the very different learning conditions in legacy apartheid schools or the linguistic diversity of the population, nor was there sufficient consideration of how to remedy the historical inequalities and damaging effects of apartheid.[6] In the enormous task of running the schooling system while changing it, the government's pre-eminent concern was to maintain system legitimacy above innovation.

Though the Constitution had declared education to be a basic right and outlawed discrimination on the basis of race, this did not translate into equality of provision for all. Establishing a single national education department and a single set of norms and standards for all schools established a mantra of 'sameness' across the system, but glossed over significant differences. New provincial mappings had no particular educational logic, resulting in a distribution of schools whose current performance largely reflects their legacy apartheid departments. The mantra of sameness does not give sufficient consideration to the historically unequal circumstances of schooling or the destructive effects of apartheid racial denigration.

With regard to funding, two aspects of post-apartheid arrangements have limited the possibilities of achieving equity: the amount of earmarked equity funding is not sufficient for improving the conditions of schools in poor communities, leaving them trapped in historical inequalities; and the introduction of fees into the public system results in major resource differentials between schools, protecting historical privilege, but also catering to new black elites. Schools where state provision is supplemented by private contributions and fees provide the model for 'good schools' in the post-apartheid imaginary as the hegemonic norm for all to aspire to. However, they are a small minority of schools, funded at a level that is not available to all. Geography traps the students of rural schools where they are, while in urban areas, mobility is viewed as the route to a better school – from former African township schools to former Coloured and Indian schools, and from those schools to former white schools. And within former white schools, a competitive market of desirability operates, as is well laid out in Mark Hunter's (2019) work on schools in Durban.

Poverty, inequality, and unemployment are structural features of the post-apartheid dispensation, which is deeply marked by inequalities of race and gender (as described in Chapter 6). South Africa is recognised as one of the most unequal countries in the world (World Bank, 2018) and the structure and performance of its schooling system contributes to this. Certainly, there will always be schools in poor communities that perform 'against the odds' and individuals in poor schools who achieve excellent results. However, these are exceptions (see Christie, Butler & Potterton, 2007). Repeating this point: the majority of the country's schools are black schools in poor communities where African languages are spoken, and the performance of these schools relative to their privileged counterparts has remained fundamentally unequal over the years.

Moreover, schools are linked to social inequalities in ways that go beyond funding and provision, as the work of French sociologist Pierre Bourdieu (1976) (among others) has highlighted. Without reading Bourdieu's analysis of schools and social reproduction too deterministically, his work points to the different relationships between families, schools, and culture. He argues that French schools match the cultural capital and ethos of the middle class, whose success at meeting the demands of the curriculum is not recognised as the 'social gift' that it is. Working class children without this cultural capital are disadvantaged from the start. Bourdieu elaborates as follows:

> In fact, to penalize the underprivileged and favour the most privileged, the school has only to neglect, in its teaching methods and techniques and its criteria when making academic judgements, to take into account the cultural inequalities between children of different social classes. In other words, by treating all pupils, however unequal they may be in reality, as equal in rights and duties, the educational system is led to give its *de facto* sanction to initial cultural inequalities.
>
> (p. 113)

In short, inequalities in schooling indicate that the power dynamics of colonialism have not substantially shifted in the decades since the decolonising moment of political change in 1994. The structure of the current system limits its capacity to redress past injustices and to produce equal experiences and outcomes for all students. The implicit assumption that all public schools will be able, with sufficient effort and 'quality control' improvement measures, to progress towards providing the same quality of education as that provided in the well-performing and well-resourced parts of the system is not realistic. It is a myth of progress which distracts from the important tasks of repair and rebuilding at the moment of decolonial change.

Border tensions

Policies for post-apartheid schooling in South Africa are pulled in competing directions: the desire to preserve excellence as defined in terms of former white schooling; and the desire to provide quality education for all, across the vastly different living conditions in the country. These competing desires bring about particular tensions as the knowledges and subjectivities of the colonial past come up against the limits of what can be achieved in the coloniality present.

In the case of Carnarvon, one simplifying feature is the fact that all townspeople speak the same language, Afrikaans (even though defined in different racial and class categories). Afrikaans is one of only two languages (along with English) that the official curriculum supports from Grade 4 onwards, despite the fact that there are 11 constitutionally recognised national languages. On a national scale, adding to the structural inequalities in the resourcing of the system is the unmediated Eurocentric orientation of post-apartheid curriculum and language policies, and their non-acknowledgement, if not silencing, of the damage done by apartheid education. Curriculum policies take for granted that the 'powerful knowledge' of the Western episteme is appropriate for post-apartheid South African students, that it is equally accessible in all schools, and that it should be delivered in the medium of English or Afrikaans from Grade 4 onwards. This approach holds sway, seemingly without question, even though most students fail against the demands of the current curriculum. Haunted by the explicit racism and discrimination of the apartheid past, post-apartheid curriculum and assessment policies are intent on emphasising 'sameness' over difference.[7] They do not acknowledge the situatedness of the curriculum knowledge they propose and the partialities and exclusions of its languages. Nor do these policies take into account the harm done by apartheid and what repair might be needed. Western superiority is naturalised and remains unquestioned – a form of 'sanctioned ignorance', to use Gayatri Spivak's (1990) term. The curriculum does not mediate its Eurocentric bias, seemingly assuming that this is the only path towards critical thinking and meaningful learning. It does not engage with local conditions and the different circumstances in which it is implemented; it does not value cultural knowledges; and it does not support the use of African languages to build on the linguistic diversity of the country (see McKinney, 2017). Neville Alexander labels the monolingual habitus of South African schooling as 'lamentable shortsightedness' (2013, p. 108) – and for good reason.

Expanding on this point, the link between language and colonial domination is powerfully articulated by Ngũgĩ wa Thiongo (1986). As an established Kenyan author publishing in English, Ngũgĩ switched his writing to Kikuyu, reflecting as follows on the destructive effects of language suppression under colonialism:

> When people or anybody alienates you from your own language, it's a kind of alienation really from many things. First, from the knowledge carried by that language – so the knowledge of the area, the trees, the rivers, whatever, gone. Second, from the history of the community that made that language, gone. For a language it takes many years, hundreds of years to be where it is. So, that external power completely whips out, like a hurricane, like a bomb that comes and clears everything that was there and it tries to plant something else on this terrain or contaminated ground, that's how I call it.
> (Barison, Carmello, El Hansali, & Pratali Maffei, 2018, p. 276)

As Ngũgĩ (1986) points out, language is more than a means of communication; it is also a carrier of culture and meaning: 'mediating between me and my own self; between my own self and other selves; between me and nature' (p. 15). In an observation that would apply to post-apartheid schooling for students speaking African languages, he states:

> Colonial alienation ... starts with a deliberate disassociation of the language of conceptualisation, of thinking, of formal education, of mental development, from the language of daily interaction in the home and in the community. It is like separating the mind from the body so that they are occupying two unrelated linguistic spheres in the same person. (p. 28)

Summing up, post-apartheid schooling with its distorted resource distribution is caught in the polarising border dynamics of coloniality. The schooling system, designed to match the institutional framework and curriculum of unmediated Eurocentric modernism, cannot simultaneously achieve its goals of redressing past injustices and providing education of progressively high quality for all. The dynamics of border conditions – the desire for what cannot be achieved, and the unwillingness to accept what exists – trap the system in its bimodal achievement patterns and its continuing inequalities. This is a cycle where the decolonisation of schooling remains an impossible dream.

Stepping out of this dynamic requires a different approach to working within border conditions. It requires a willingness to acknowledge the ways in which existing intersectional inequalities and power relations are reflected in schooling, so as to work against their inevitabilities and redress their consequences in a spirit of repair. It requires a willingness to shift the structure of resource allocations to schooling so that historical inequalities are adequately redressed. It requires a willingness to engage with the actual conditions of learning of the majority of students in order to build an education, which, in Paolo Freire's terms, replaces a 'banking' approach with the development of critical capacities to read the world in order to change it. It requires recognition of languages spoken by students and the development of these as languages of formal education. If the goals of justice, redress, and quality education for all are to be given serious weight, it is necessary to review current practices, including funding, curriculum, and language policies, and be willing to change them as needed. In Freire's words, 'Education is an act of love, and thus an act of courage. It cannot fear the analysis of reality or, under pain of revealing itself as a farce, avoid creative discussion' (1974/2005, p. 33).

In short, what is required is a different ethical imagination informing the provision of an education which values intellectual rigour as well as care for others and a common good, and a political commitment to working collectively to achieve this. In the sections that follow, I explore the ethical basis to support a redesigned system that would affirm the equal dignity and value of all in relationships grounded in respect, reciprocity, and care.

Towards an 'all-world' ethics

The conceptual challenges of current times entail grappling with the legacies of colonialism and imperialism alongside additional challenges: the effects of global neoliberalism; developments in technology and their implications; violence-induced mobilities and xenophobia; ecological crises and climate change; and issues of global health pandemics including HIV/Aids and COVID-19. The confluence of these imperatives calls for new ways of understanding and living in the world beyond existing ontologies of difference and also including the more-than-human world. I suggest that this requires reclaiming and reconfiguring what it is to be human, towards developing a different approach to ontology, ethics, and politics.

In rethinking the human subject in current times, it seems appropriate to return to Fanon as a starting point, given his powerful exhortation for humanity beyond the devastation and trauma of colonisation, the violence of ending it, and the disappointments of an inadequate national bourgeoisie. Fanon's legacy, though dated in some ways, still speaks powerfully to the decolonial task of reconfiguring a common humanity. In this regard, Achille Mbembe (2017) has observed:

> If we owe Fanon a debt, it is for the idea that in every human subject there is something indomitable and fundamentally intangible that no domination – no matter what form it takes – can eliminate, contain or suppress, at least not completely. Fanon tried to grasp how this could be reanimated and brought back to life in a colonial context that in truth is different to ours, even if its double – institutional racism – remains our own beast. For this reason, his work represents a kind of fibrous lignite, a weapon of steel, for the oppressed today.
>
> (p. 170)

In seeking new directions for life beyond colonisation, Fanon (1961/1990) was deeply cynical about what the European version of humanism could offer: that for all its proclamations about 'man' it was murderous in its actions and stifling of humanity across the globe; it stratified the world; and it was incapable of inclusivity and collaboration with others. Fanon's call for action was a call 'to work out new concepts', a new history beyond Eurocentric formulations. These points remain applicable today. In this spirit, he famously stated that 'Each generation must, out of relative obscurity, discover its mission, fulfil it or betray it' (p. 166).

With regard to legacies of colonialism and its ontologies of difference, race remains an active signifier that must be tackled.[8] Important postcolonial and decolonial theorists have sought to unsettle the notion of race and to foreground its configuration in the systemic injustices of western domination.[9] In *Critique of Black Reason*, Mbembe (2017) recognises there does not yet exist a 'world-beyond-race', and that as long as racism persists in imagination and practice, the struggle to create a different world will be necessary. However, he proposes conjuring with the signifier 'black' and extending it, 'in order to reaffirm the innate dignity of every human being and of the very idea of a common humanity, a same humanity, an essential human resemblance and proximity' (p. 173). Recognising that the scars of history require restitution and reparation, Mbembe calls for a position beyond victimhood towards building a shared world of equal dignity and care for the other in a common humanity:[10]

> To build a world that we share, we must restore the humanity stolen from those who have historically been subjected to processes of abstraction and objectification. From this perspective, the concept of reparation is not only an economic project but also a process of reassembling amputated parts, repairing broken links, relaunching the forms of reciprocity without which there can be no progress for humanity.
>
> (p. 182)

In South Africa, where shifting configurations of structural racism and race-thinking have prevailed over time and continue into the present, there can be no easy route to nonracialism or a world beyond race. The task of repair cannot be simplified into a meme of sameness, as current education policies assume, without the accompanying labour of recognition, revaluing, and reconstruction. The remaking of social and economic relationships beyond colonialism, including schooling, requires ontological as well as political work.

For Mbembe, reparation requires expanded concepts of justice, dignity, reciprocity, and concern for the other – extending beyond the human:

> There is only one world. We are all entitled to it by the fact of our very existence. The only way in which to ensure its sustainability or its duration is to share it as equitably as possible.
>
> And when I say we must share it, I do not simply have humans in mind; we must share it with every other existent, and in so doing, reinvent democracy.
>
> (Goldberg, 2018, p. 217)

The work of Mbembe and Fanon as well as others calls for the reconfiguration of notions of the human and humanity, and the relations we establish with others in a shared world.[11] Specifically, this reconfiguration requires a shift from Enlightenment thinking, its ways of dividing the world, and the central position it accords to the concept of 'Man' as separate from and superior to the rest of the world and other beings.

Ontological and ethical reframing

Reconfiguring the notions of human and humanity requires relational rethinking about ontology, about the nature of the world in its form and substance, in order to include not only humans and other living beings but also the earth itself as the basis of life. The earth has its own forces of change, including geological changes, changes in climate and ecology, and planetary changes. As theorists such as Kathryn Yusoff (Yusoff, 2015; Yusoff, Grosz, Clark, Nash & Saldanha, 2012) have proposed, the notion of geopower (the power of the earth) needs to be considered alongside biopower (power operating through life and the body) as part of the substance of the world. Ontological reframing of this order to include the earth and its powers as well as all forms of life, human and more-than-human, requires an accompanying reconsideration of ethics – how we might best live together with all others and the earth. And, as Liz Grosz's project on ontoethics shows, a reconfiguring of ontology and ethics also involves politics 'which addresses social, collective, cultural, and economic life and their possibilities for change' (2017, p. 1).

If ontology, ethics and politics are understood to include all-world considerations beyond the human and including the earth itself, the nature of climate change also needs fuller consideration. The ways in which climate change is to be understood, and the implications of theorising this in the 'Anthropocene narrative',[12] is the subject of much continuing debate, the detail of which lies beyond the scope of this chapter. Also beyond the scope of consideration here is the global COVID-19 health pandemic, unfolding as this book goes to press and highlighting the shared vulnerability of all human life.

Whatever the understanding on issues of climate change and global health, it is increasingly clear that important shifts in understandings of humanity and the earth are taking place. These shifts acknowledge the limits of human knowledge and the need to embrace uncertainty and complexity. I suggest that for education, these shifts require a relational ethics of continuously assessing how humans might best live together with others and the earth itself – a shift I refer to as the relational ethics of the all-world – and building schools that sustain this ethics.

Schooling and the all-world

Running through the chapters of this book is the theme of schooling as a significant social institution, with its mandate of providing systematic teaching and learning for young people, hopefully in ways that will prepare them to contribute to a shared world and change it for the better. Entangled as schools are in broader political economies and social relations, they have limited possibilities for leading change, being far more likely to reproduce their broader contexts than to challenge them. That said, there are always opportunities to shift current practices in and beyond schools to improve the quality of learning experiences for all students. These opportunities for improvement – be they first-order changes in quality control or second-order changes in design – must always be worked with, both within and beyond formal schooling, towards the telos of building a shared world.

In the post-apartheid period, much of the emphasis on education has rested on formal structures and government policies and actions. The dominant narrative is government-centred, and, though this is important, it should not be regarded as the only space for action. It should not overshadow South Africa's history of resistance and radical alternatives in education, both within and outside of schools.[13] As Salim Vally reminds us:

> South Africa has a proud legacy of education for liberation comprising a history of resistance in and through education. This resistance generated various popular epistemologies and pedagogies including the 'peoples' education movement', 'worker education' the 'popular adult and/ or community education movement' and 'education with production'. There were also many community-based initiatives around early childhood development, reading and literacy.
>
> (2019, p. 2)

South Africa's legacy of political and social activism around education provides a reservoir of experience for critical reflection and action in decolonising projects in schooling, alongside broader social, political, and economic movements for change. For it must be remembered that colonialism and decolonisation in South Africa are based on capitalism, currently configured in the markets, consumerism, and hyper-individualism of neoliberalism. Under these conditions, struggles against inequalities involve struggles of economic ownership and distribution in various significant intersectional configurations.

There are no blueprints for a decolonial schooling system. What is required is to work beyond the inevitabilities of current inequalities towards a different imaginary and different aspirations for schooling, as well as for education beyond schooling. This brings with it an ethical stance of reciprocal, collective responsibility and repair, towards affirming human dignity and equal sharing in one world. And it requires a political commitment to the collective work of social and economic change, within which schooling plays its part.

Beyond victimhood and guilt, beyond blame and recrimination, lie the possibilities of building the all-world that we share with others, human and more-than-human, and the earth itself.

Notes

1 As a Marxist humanist, Fanon placed little hope in the national bourgeoisie to work for the betterment of the majority:

> In under-developed countries … no true bourgeoisie exist; there exists only a sort of little greedy caste, avid and voracious, with the mind of a huckster, only too glad to accept the dividends that the former colonial power

> hands out to it. This get-rich-quick middle class shows itself incapable of great ideas or inventiveness. It remembers what it has read in European textbooks and imperceptibly it becomes not even the replica of Europe, but its caricature.
>
> (p. 141)

While the racial prejudice of the Western bourgeoisie is one of contempt towards 'the nigger and the Arab', the racial prejudice of the national bourgeoisie 'is a racism of defence, based on fear' (p. 131).

2 In an IMF Working Paper, Mlachila and Moeletsi (2019) sum up the position as follows:

> In a group of fourth-grade learners from 49 countries, South African fifth-grade learners ranked bottom last in Trends in International Mathematics and Science Study (TIMSS) test scores. Moreover, secondary education is failing to address these gaps. In a group of 39 mathematics and science eighth-grade learners, South Africa's ninth-grade learners rank bottom and second last, respectively, in TIMSS test scores. Peer countries such as Kenya, Swaziland, and Botswana out-rank South African learners in reading and mathematics scores.
>
> (p. 5)

3 Mlachila and Moeletsi, p. 5: 'The poorest 75–80 percent of learners depend on dysfunctional public schooling and achieve poor outcomes while wealthiest 20–25 percent of learners enrol in private schools and functional public schools, and achieve better academic outcomes'.

4 Of course they were not alone in securing these existing privileges but with regard to schools, their determination was crucial in securing the constitutional settlement, as described in Chapter 7. All former white schools were 'protected' through these measures.

5 It is interesting to note the very formal nature of government processes of 'consultation', for example through invitations to comment in formal governmental announcements, that assume a literate population that reads this form of documentation and responds in writing to it – surely a small minority of the population. This process stands in strong contrast to the traditions of consultation evident in the anti-apartheid struggle.

6 On acknowledging the psychological damages of colonialism and colonial education, see also Abdi (2011); Biko (1996); Memmi (1965); Nandy (2009); Rodney (1982).

7 It is worth noting the problematic history of post-apartheid curriculum, from the outcomes-based Curriculum 2005, through its revisions, to the current Curriculum and Assessment Policy Statements (CAPS).

8 I acknowledge that the account I have provided does not adequately consider other intersectional inequalities, particularly gender, and including sexuality and religion.

9 There is an enormous and significant body of work by anti-colonial and postcolonial scholars over decades that this research acknowledges but does not cite for reasons of space.

10 The move beyond victimhood, which Fanon also calls for, is a complex shift to propose, particularly given the persistence of racism across in the world beyond the formal end of colonialism. Fanon's call is for an existential act of freedom against dehumanization. Both black and white 'must turn their backs on the inhuman voices which were those of their respective ancestors in order that authentic communication be possible'. Beyond relationships of superiority and inferiority is the 'the quite simple attempt to touch the other' in a reciprocal relationship 'building the world of the You' (p. 165).

11 For an elaboration of *tout-monde*, see Glissant (1997). For reconfiguring the human, see Wynter (2003). For the politics of an alternative modernity building on difference see Cusicanqui (2012).

12 If human activity is indeed responsible for ushering in a new global epoch – the Anthropocene – then human history and natural history cannot be separated, but need to be considered together. This itself calls into question the disciplinary divides between humanities and natural sciences and how this particular moment might best be understood. Concerns such as these are articulated in Dipesh Chakrabarty's (2012) 'Four Theses' and subsequent debates e.g., Luisetti, 2019). In relation to the scale of climate change, Chakrabarty questions the adequacy of existing debates on globalisation, modernity, and the history of capitalism, and calls for a fundamental rethinking of human understanding of the world. The centrality of 'anthropos' in this narrative is challenged by alternative namings such as 'capitalocene' and 'chthulocene'.

13 My book, *The Right to Learn,* was written within an alternative education project, SACHED Trust.

References

Abdi, A. A. (2012). Decolonizing philosophies of education. In A. A. Abdi (Ed.), *Decolonizing philosophies of education* (pp.1–13). Rotterdam: Sense Publishers.

Alexander, N. (2013). *Thoughts on the new South Africa.* Johannesburg: Jacana Press.

Barison, G., Carmello, B., El Hansali, A., Pratali Maffei, D. (2018). Ngũgĩ wa Thiong'o: An Interview. *Il Tolomeo*, 20 December.

Biko, S. (1996). *I write what I like.* Johannesburg: Ravan Press.

Bourdieu, P. (1976). The school as a conservative force: Scholastic and cultural inequalities. In R. Dale, G. Esland, & M. MacDonald (Eds.), *Schooling and capitalism: A sociological reader* (pp. 110–117). London: Routledge.

Chakrabarty, D. (2012). The climate of history: Four theses. *Critical Inquiry*, 35(2), 197–222.

Christie, P. (1991). *The right to learn.* Johannesburg: SACHED Trust/Ravan Press. (First edition 1985)

Christie, P., Butler, D., & Potterton, M. (2007). Report of the ministerial committee on schools that work. Republic of South Africa: Department of Education.

Christie, P., & Monyokolo, M. (2018). Reflections on curriculum and education system change. In P. Christie, & M. Monyokolo (Eds.), *Learning about sustainable change in education in South Africa: The Jika iMfundo campaign, 2013–2017* (pp. 265–272). Johannesburg: Saide.

Cuban, L. (1988). A fundamental puzzle of school reform. In A. Lieberman (Ed.), *Schools as collaborative cultures: Creating the future now* (pp. 71–77). New York: Falmer Press.

Cusicanqui, S. R. (2012). *Ch'ixinakax utxiwa:* A reflection on the practices and discourses of decolonization. *South Atlantic Quarterly, 111*(1), 95–109.

Fanon, F. (1961/1990). *The wretched of the earth.* London: Penguin Books.

Fanon, F. (1970). *Black skin white masks.* London: Paladin, p. 165.

Freire, P. (1974/2005). *Education for critical consciousness.* London; New York: Continuum.

Glissant, E. (1997). *Poetics of relation.* Ann Arbor: University of Michigan Press.

Goldberg, D. T. (2018). 'The reason of unreason': Achille Mbembe and David Theo Goldberg in conversation about *Critique of Black Reason. Theory, Culture & Society, 35*(7–8), 205–227.

Grosz, L. (2017). *The incorporeal: Ontology, ethics and the limits of materialism.* New York: Columbia University Press.
Hunter, M. (2019). *Race for education: Gender, white tone, and schooling in South Africa.* Cambridge: Cambridge University Press.
Lefebvre, H. (1991). *The production of space.* Oxford: Blackwell.
Luisetti, F. (2019). Geopower: On the states of nature of late capitalism. *European Journal of Social Theory, 22*(3), 342–363.
Mbembe, A. (2017). *Critique of Black reason.* Durham: Duke University Press, p. 170.
McKinney, C. (2017). *Language and power in post-colonial schooling: Ideologies in practice.* New York: Routledge.
Memmi, A. (1965). *The colonizer and the colonized.* Boston: Beacon Press.
Mlachila, M., & Moeletsi, T. (2019). *Struggling to make the grade: A review of the causes and consequences of the weak outcomes of South Africa's education system.* IMF Working Paper WP/19/47.
Nandy, A. (2009). *The intimate enemy: Loss and recovery of self under colonialism.* Oxford: Oxford University Press.
Ngũgĩ wa Thiongo. (1986). *Decolonising the mind: The politics of language in African literature.* London: James Currey.
Rodney, W. (1982). *How Europe underdeveloped Africa.* Washington, DC: Howard University Press.
Spivak, G. C. (1999). *A critique of postcolonial reason: Toward a history of the vanishing present.* Cambridge, MA: Harvard University Press.
Spivak, G. (2004). Righting wrongs. *The South Atlantic Quarterly, 103*(2), 523–581.
Vally, S. (2019). Between the vision of yesterday and the reality of today: Forging the pedagogy of possibility. Professorial Inauguration, University of Johannesburg. 4 September 2019.
World Bank. (2018). *Overcoming poverty and inequality in South Africa: An assessment of drivers, constraints and opportunities.* Washington, DC: World Bank.
Wynter, S. (2003). Unsettling the coloniality of being/power/truth/freedom: Towards the human, after man, its overrepresentation – An argument. *CR: The New Centennial Review, 3*(3), 257–337.
Yusoff, K. (2015). Anthropogenesis: Origins and endings in the Anthropocene. *Theory, Culture & Society, 33*(2), 3–28.
Yusoff, K., Grosz, E., Clark, N., Nash, C., and Saldanha, A. (2012). Geopower: A panel on Elizabeth Grosz's Chaos, territory, art: Deleuze and the framing of the earth. *Environment & Planning D: Society & Space, 30*(6), 971–988.

Index

Note: **Bold** page numbers refer to tables; *italic* page numbers refer to figures/maps and page numbers followed by "n" denote endnotes.

For Product Safety Concerns and Information please contact our EU
representative GPSR@taylorandfrancis.com
Taylor & Francis Verlag GmbH, Kaufingerstraße 24, 80331 München, Germany

www.ingramcontent.com/pod-product-compliance
Ingram Content Group UK Ltd.
Pitfield, Milton Keynes, MK11 3LW, UK
UKHW022026061126
477531UK00002B/52
9780367425753